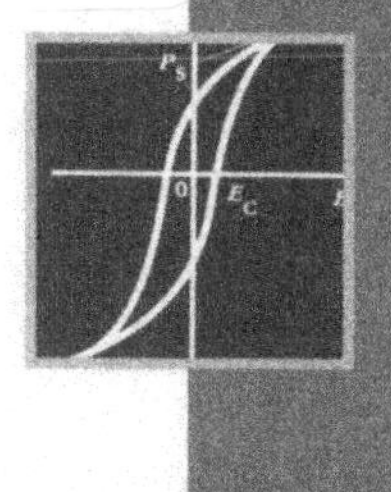

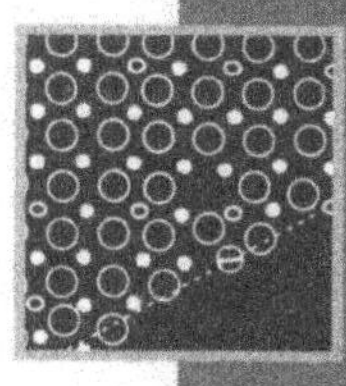

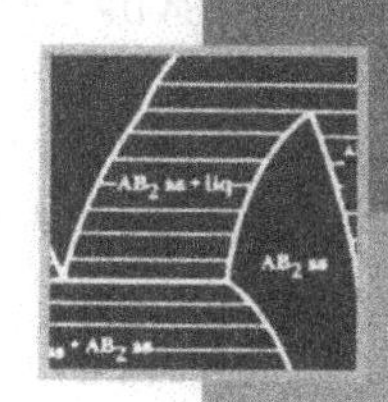

PHASE DIAGRAMS AND CERAMIC PROCESSES

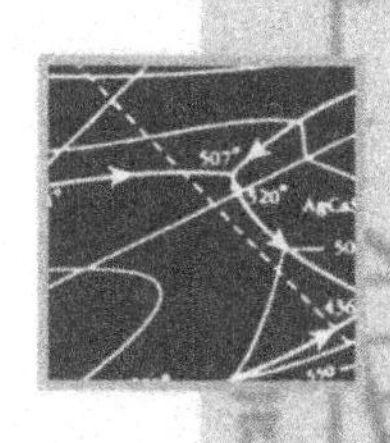

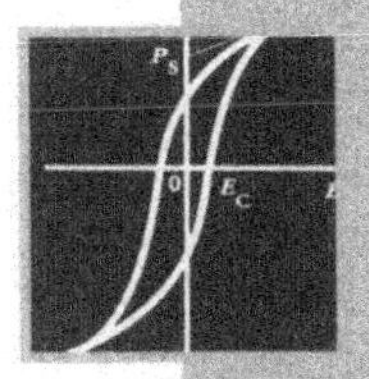

PHASE DIAGRAMS AND CERAMIC PROCESSES

Anna E. McHale

Consultant
Alfred, New York

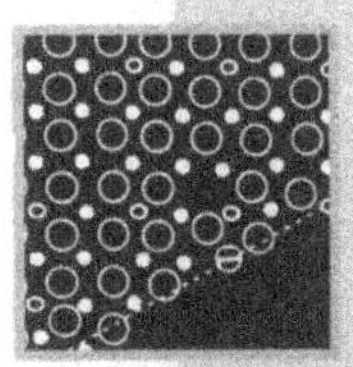

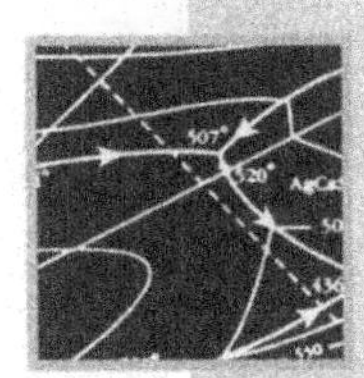

CHAPMAN & HALL

INTERNATIONAL THOMSON PUBLISHING

New York • Albany • Bonn • Boston • Cincinnati • Detroit • London • Madrid • Melbourne
Mexico City • Pacific Grove • Paris • San Francisco • Singapore • Tokyo • Toronto • Washington

Cover design: Curtis Tow Graphics

Printed in the United States of America

Chapman & Hall
115 Fifth Avenue
New York, NY 10003

Chapman & Hall
2-6 Boundary Row
London SE1 8HN
England

Thomas Nelson Australia
102 Dodds Street
South Melbourne, 3205
Victoria, Australia

Chapman & Hall GmbH
Postfach 100 263
D-69442 Weinheim
Germany

International Thomson Editores
Campos Eliseos 385, Piso 7
Col. Polanco
11560 Mexico D.F
Mexico

International Thomson Publishing–Japan
Hirakawacho-cho Kyowa Building, 3F
1-2-1 Hirakawacho-cho
Chiyoda-ku, 102 Tokyo
Japan

International Thomson Publishing Asia
221 Henderson Road #05-10
Henderson Building
Singapore 0315

1 2 3 4 5 6 7 8 9 10 XXX 01 00 99 98 97

Library of Congress Cataloging-in-Publication Data

McHale, Anna.
Phase diagrams and ceramic processes / Anna McHale.
p. cm.
Includes bibliographical references and index.

1. Ceramics. 2. Ceramic materials. 3. Phase diagrams.
I. Title.
TP810.5.M35 1997
666--dc20 96-38438
CIP

ISBN 978-1-4419-4726-0

British Library Cataloguing in Publication Data available

To order this or any other Chapman & Hall book, please contact **International Thomson Publishing, 7625 Empire Drive, Florence, KY 41042.** Phone: (606) 525-6600 or 1-800-842-3636.
Fax: (606) 525-7778. e-mail: order@chaphall.com.

For a complete listing of Chapman & Hall titles, send your request to **Chapman & Hall, Dept. BC, 115 Fifth Avenue, New York, NY 10003.**

Dedicated to the memory of my Father, an engineer who cared

Contents

Preface

Ceramic products are fabricated from selected and consolidated raw materials through the application of thermal and mechanical energy. The complex connections between thermodynamics, chemical equilibria, fabrication processes, phase development, and ceramic properties define the undergraduate curriculum in Ceramic Science and Ceramic Engineering.

Phase diagrams are usually introduced into the engineering curriculum during the study of physical chemistry, prior to specialization into ceramic engineering. This creates an artificial separation between consideration of the equilibrium description of the chemically heterogeneous system and the engineering and physical processes required for phase, microstructure, and property development in ceramic materials. Although convenient for instructional purposes, the separation of these topics limits the effective application of phase diagram information by the ceramic engineer in research and manufacturing problem solving.

The nature of oxide phases, which define their useful engineering properties, are seldom linked to the stability of those phases which underlies their reliability as engineered products. Similarly, ceramic fabrication processes are seldom discussed within the context of the equilibrium or metastable phase diagram.

In this text, phase diagrams are presented with a discussion of ceramics' properties and processing. Particular emphasis is placed on the nature of the oxides themselves—their structural and dielectric properties—which results in unique and stable product performance.

Any set of systematic property measurements can be the basis for a phase diagram: every experiment is an experiment in the approach to phase equilibrium. An engineer with a good working understanding of the approach to equilibrium, equilibrium phase diagrams, and kinetic processes has the tools at hand for effective materials engineering. This book should provide a framework within which the "conventional wisdom" of the industry or personal observations can be quantified and extended for application to new systems.

The topical discussion is organized in a manner suitable for use in a semester course for undergraduates who have completed basic course work in physical chemistry, including a brief introduction to phase equilibrium diagrams. The student should be familiar with ceramic raw materials and have some experience of materials characterization and the conventional processing of crystalline and amorphous materials.

Anna E. McHale
Alfred Station, NY
1997

Acknowledgments

This manuscript grew from my needs for teaching materials for undergraduate ceramic engineers in the applications of phase equilibrium diagrams, and was researched and prepared using the facilities and resources of the New York State College of Ceramics at Alfred University, Alfred, NY. The writing of this book would have been impossible without the continuing and loving support of my husband and family.

The help of the Ceramic Engineering and Materials Science Faculty and the Staff of the Scholes Library of Ceramics is gratefully acknowledged.

1

Introduction

From the beginnings of civilization, ceramics have been prized for their strength, durability, and beauty. The first "high tech" ceramics, possessing advanced properties produced by systematic experimentation, are found as early as 5000 BC in Egyptian faience and Chinese glazed pottery. Ceramic containments for metal smelting and glass melting were in use before 2000 BC. Hydraulic cements and brick technologies enabled the building of the great structures and roads that characterized the expansion of the Roman Empire. Throughout recorded civilization, technological advances in all fields have been predicated on or facilitated by advances in ceramic science and technology.

Until the 1700s, only incremental and nonsystematic progress in ceramic technology was made, as there existed minimal communication between practitioners and no means of analysis of either raw materials or produced ware. Application of the new sciences of mineralogy and chemistry brought about rapid improvement in ceramic quality during that century and a revolution in the understanding of materials came with the invention of practical metallography in the mid 1800s. With this technological advance, the multiphase nature of metals and ceramics became apparent.

During this same period, thermodynamics and physical chemistry became established and the characteristics of the chemical elements were systematized into the modern periodic table. In 1877, Gibbs' work into the nature of heterogeneous systems produced the *phase rule,* linking the thermodynamic descriptions of phases and phase composition with the conditions of phase equilibrium.

The Gibbs free energy, or the energy available for phase change, is defined as $\Delta G = \Delta H - T\Delta S$, a generalization of the condition for isothermal thermomechanical equilibrium $\Delta H = T\Delta S$, the balance between system enthalpy and entropy. Gibbs showed that in a multicomponent, multiphase system, the equivalence of the partial differential of ΔG with respect to each component's concentration in each phase was the essential requirement for equilibrium between phases. The partial differential quantity was defined as the chemical potential,

$$\mu_i = \left(\frac{\partial \Delta G}{\partial x_i}\right)_{T,P,x_j}.$$

The number of phases of a multicomponent system that could coexist under equilibrium conditions could be determined using the Gibbs phase rule.

While Gibbs' work required no assumptions on the nature of matter or chemical species, the demonstration of practical X-ray diffraction crystallography early in this century proved the atomic nature of matter and the uniqueness of crystalline and amorphous phase structures. Chemistry and periodicity were thereafter firmly linked in the description of phases and their properties.

The modern ceramist's ability to characterize raw materials and their reactions, to understand the atomistic development of phases and microstructures, and to relate this information to the performance of ceramic products is a very recent development in the over 7000-year history of ceramic technology. The continued developments of materials science and solid-state physics have inspired development of entirely new classes of oxide and nonoxide ceramic materials for refractory, electronic, and magnetic application. The phase diagram has become the essential means for rapid and precise communication of the chemical equilibrium data derived from the constant search for improved ceramic products and processes.

A phase diagram is simply a compact presentation of the known or postulated equation of state of a closed chemical system. The equation of state is a relationship between the state variables—pressure, temperature, and system composition—and the macroscopic equilibrium of the system and its phases and phase proportions. A phase diagram is a convenient means to present a known equilibrium relationship (either experimentally or theoretically derived), extend a known relationship or equation of state to as-yet-unexplored experimental conditions, or to postulate equilibrium phase relationships from limited observations. The ability to use the phase diagram format to extend or postulate equilibrium phase relations derives from the general regularity seen in the thermodynamic functions of both pure phases and solutions.

The equilibrium phase diagram can also be used to postulate reasonable non-equilibrium or metastable states obtained as a result of characteristic system kinetics or through manipulation of process factors. Metastability describes a state with a calculable energy separation from equilibrium; that is, the energetic states of the phases present may be extrapolated from equilibrium values.

When a system is "quenched"—rapidly cooled to prevent equilibrium processes from proceeding owing to sluggish kinetics—a "stranded" condition is obtained in which the energetic condition of the phases can no longer be determined through extrapolation from the equilibrium condition. For the stranded system, the equilibrium phase diagram contains valuable information on the gross separation from equilibrium, the nature of plausible reaction processes that would reduce the energetic separation from equilibrium, and the required kinetic or thermal energy that might reasonably be required for equilibration to proceed.

1.1 Ceramic Phases and Phase Equilibria

Metallurgical phase diagrams are astounding in their detailed and precise relationship to microstructural development in steels and alloys.[1] The oxide phase diagram, on the other hand, is often less precise and generally more difficult to determine from microstructural observations. This difference reflects a basic crystal chemical difference between metallic and ionic materials and their related defect and transport properties. The stability and durability exhibited by oxide ceramics is a direct reflection of crystallographic structure and stoichiometry requirements of ionic bonding.

The oxide structure is composed of at least two sublattices, anionic and cationic. These are related through symmetry, reflecting overall charge balance and the directional nature of the chemical bonding—a mixture of covalent and ionic types. The higher the degree of ionic character present in the bonding, the greater the strength of the directional bond. This is reflected in the high melting points and high stiffness of the crystal lattice.

Additional sublattices can be defined that describe related structural elements typical of the lattice and specific to chemistry and temperature. In Al_2O_3, for example, the O^{2-} sublattice is hexagonal close-packed. The Al^{3+} occupy two thirds of the interstitial spaces of the array, yielding overall electroneutrality. The occupied sites define the second sublattice. The third sublattice is defined by the one third of sites not normally occupied by Al^{3+}.

The energetic description of the oxide and its equilibrium condition reflects the necessity to maintain overall and point-to-point charge neutrality and crystal symmetry, linking the three sublattices. In practical terms, correlation requires that a chemical alteration on one sublattice (as preceding reaction or nucleation of a phase change) requires a related change on the other sublattices. The net effect in Al_2O_3 and other refractory oxides is the observed relatively inert and kinetically sluggish behavior of refractory oxides and related mineral phases, particularly against solid state reactions at low and moderate temperatures.

1.2 Oxides in the Environment

Studies of natural oxide mineral phases, their occurrence, and the nature of long-term interactions with the environment have led to great advances in the

---

[1]Ceramic engineers can be extremely proud of their enormous contribution to this valuable data base, as its development would have been impossible without reliable refractory oxide containments, stable insulation, abrasives, and precision optics for metallography.

understanding of complex oxide thermochemistry and phase equilibria. In these stable systems, partial equilibration of the originally stranded phases with their environment can be observed. Banded structures, solute partitioning, recrystallization phenomena, and long-term corrosion processes reflect the influence of thermal, stress, and potential gradients acting over millions of years. Geological upheavals have also led to sudden pressure changes in rock strata, impossible to duplicate in the laboratory. Through studies of geological systems, the occurrence of stable compounds has been generalized to the natural tendency toward reaction and compound formation between oxides of cations with highly different ionization potentials (see Chapter 8).

Geological studies have been essential in providing an understanding of solute and defect behaviors in complex oxides. Geological interaction of oxide minerals with environmental water and dissolved cations is especially instructive to the engineer concerned with ceramic-water interactions. Cation or anion incorporation at an interface is immediately balanced by adsorption or incorporation of OH^- or H^+. Feldspars, microcracked as a result of thermal and baric stress, are gradually altered in this way to form hydrated clays. The progress of the reaction is facilitated by the relatively open sublattice structure of the silicate mineral which enables diffusion of exchanged cations on one sublattice to be balanced by percolation of protons (loosely bound as OH^-) in the oxygen sublattice. The exchange of mass and charge, with the lack of a distinct nucleation step, classifies weathering of silicates as a corrosion reaction.

Geological study has led to an understanding of defect processes in reaction. Corrosion reaction is facilitated by the existence of an empty sublattice in the original structure and a large surface area for interaction. Close-packing on all sublattices, as is found in rocksalt structures (MgO) and perovskites, does not allow ready solid-state corrosion. In these structures, interaction must occur by nucleated surface reaction or through dissolution of the surface with reprecipitation.

All geological and progressive interaction reflects the thermodynamic tendency toward reduced free energy. In an open system (mass not conserved) the direction of equilibrium reaction is determined through comparison of formation energies of the possible phases which could be formed from the elements present. In a closed system (mass conserved), equilibration processes reflect the need for minimization of the chemical potential difference for each species in each phase present which must be the same in each phase at equilibrium. In oxide systems, the Gibbs' energy associated with chemical change may include large contributions from entropy, enthalpy, and internal energy associated with lattice defect formation and local bonding changes. The stoichiometric requirements of overall charge neutrality and lattice site conservation are physically related to the energy balance reflected in the chemical potential.

Reasonable criteria for stability such as persistence and apparent nonreaction are not equivalent to the criteria for equilibrium. In the stable system, available kinetic energy is insufficient to allow the measurable approach to equilibrium.

Available kinetic energy is quantified in the absolute temperature; each atom in a condensed phase possesses $\approx k_b T$ of available vibrational or kinetic energy.

The relative sufficiency of the available energy versus the energy required for complete component reorganization, as required for equilibration, is judged by the "homologous" temperature scale, $T_h = T/T_m$ for each phase. A stranded phase is generally stable at $T_h \leq 0.6$ but will show increasing tendencies for progressive equilibration at higher relative temperatures. Typically high energy requirements for equilibration in oxide systems are responsible for the remarkable phase stability that is the hallmark of a ceramic product.

1.3 Phase Diagrams and Ceramic Processes

The creation of useful ceramic products involves numerous process stages. At each stage, the engineer makes decisions based both on practicality and science. Raw materials must be bought from a cost-effective source and must also be appropriate for desired reaction. Additives and means of consolidation are chosen to promote homogeneity and uniformity, as flaws in either lead to degraded product or failure during manufacture. Each chemical species introduced into the system, intentionally or unintentionally, is potentially important as a thermodynamic component. The component assembly sequence and methods, the means of consolidation, the degree of phase contact, and overall mixing homogeneity will influence the initiation and course of reaction processes at later stages.

In Figure 1.1 the steps typical in ceramics processing are diagrammed. At the left are typical process decisions at each stage. At the right, thermodynamic characteristics that are affected by these decisions are noted.

Mechanical, electromagnetic, thermal, and corrosive environmental conditions will affect the stability of the product in use.

Interaction with a service environment is idealized in Figure 1.2. The ceramic product, usually fabricated by sintering, possesses mechanical, electrical, and thermal properties that relate to the nature and microstructural relationships of the individual phases. The ceramic performance is that of a composite structure that, over time, reflects the thermodynamic and kinetic stability of that structure as predicted by the phase equilibria of the ceramic and its environment.

The study of ceramic processing and properties cannot be separated from the processes of reaction and phase equilibria. In this book, the student will find a discussion of oxide properties related to lattice structure and an introduction to the defect and transport properties characteristic of oxides. Defects in oxides are related directly to the chemical and phase stability of ceramic systems.

Basic thermodynamic relationships that are specifically useful in the description of multicomponent, multiphase equilibria and phase diagrams are reviewed. The discussion includes binary and ternary equilibrium and kinetically limited crystallization behaviors. The final discussion relates reaction behavior during ceramic processing and application to phase equilibrium data.

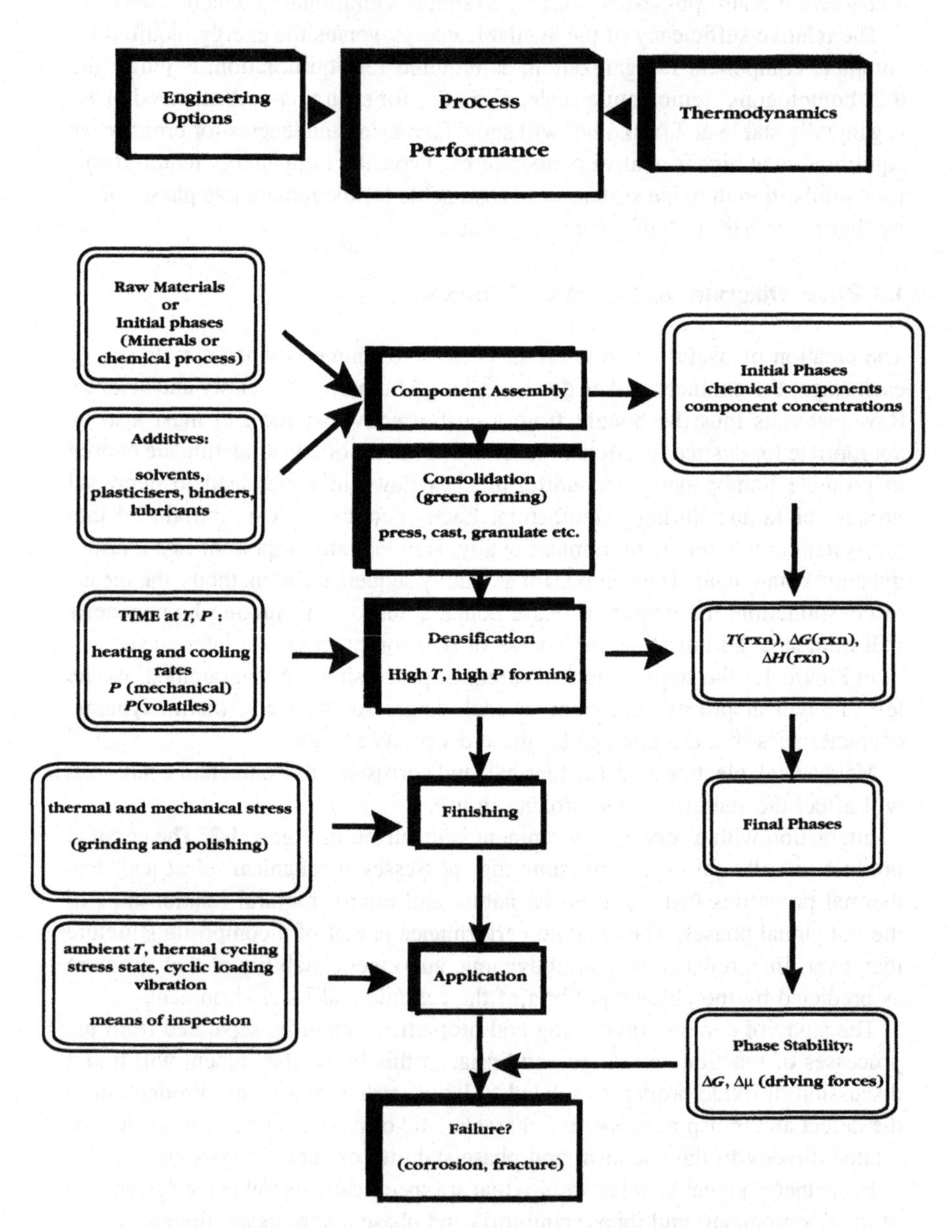

Figure 1.1 Typical materials processing steps in ceramic fabrication.

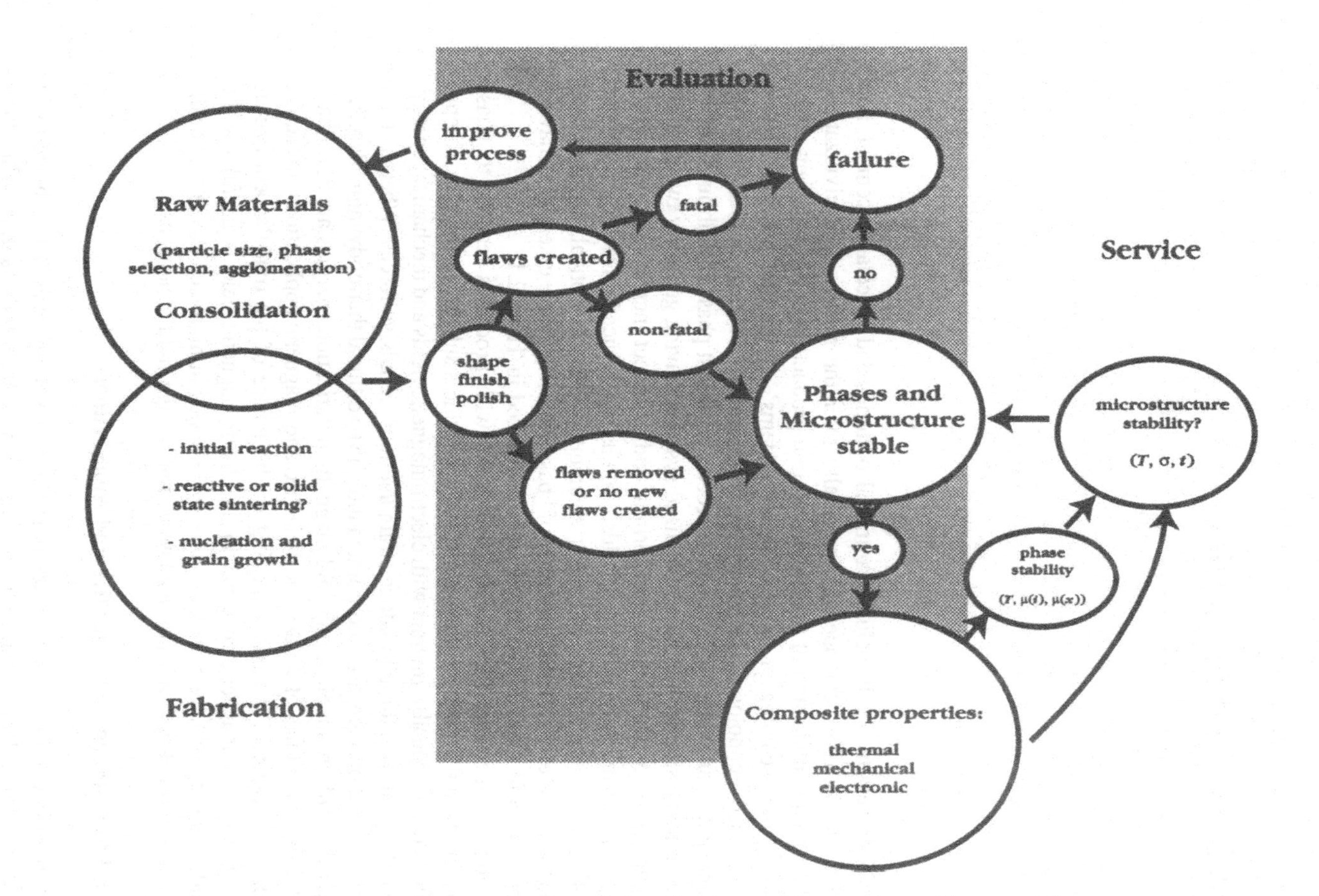

Figure 1.2 Cyclical aspects of environmental interaction with materials.

2

Products and Properties

Oxide ceramics still include traditional potteries and porcelains, cement, and brick which have developed incrementally over the millennia of civilization. Widespread application of traditional ceramic products does not represent widespread understanding of these complex systems' equilibrium behavior. These established technologies are among the most complex in terms of phase equilibrium studies and their applications in processing. Most phase equilibrium studies are on much simpler and rather idealized systems, having generally two or three components. Modern phase equilibrium studies reflect increasingly demanding applications in many traditional fields, such as in advanced refractories, and late 20th century applications of oxides as unique dielectric, optical, acoustic, and magnetic materials. A small number of "base" materials have emerged as having application in many areas. These are summarized in Table 2.1.

High temperatures are usually associated with processes leading to phase instability and possible failure. However, the oxide's sublattice structure of separated charges can also interact with electromagnetic fields and mechanical energy, readily quantified through lattice strain and ionic repulsive forces. Whereas temperature and pressure are incorporated into the traditional thermodynamic description of equilibrium, electronic and magnetic polarization energies are not. Phase equilibrium studies of oxides for dielectric and magnetic applications are active fields of research. An understanding of the nature of the oxide's behavior as a dielectric or magnetic material in an electromagnetic field is useful for the understanding of long-term behaviors of ceramics in high-tech devices or traditional electrical and thermal insulations under high field conditions.

2.1 Electromagnetic Properties and Applications

When placed in an electromagnetic field, the largely ionic character of the periodic oxide structure defines stable charge separation, quantified through the Madelung

Table 2.1 Summary of Oxides in Common and Not So Common Applications

Oxide or Compound	Type	Applications
Al_2O_3	α	Refractory thermal insulations (monolith, castables, blanket and reticulated forms) Refractory structures Electrical insulation Transparent optics Abrasives Catalyst supports Chemical filtration media
Al_2O_3	β,($Na_2O{\cdot}11Al_2O_3$)	Refractory forms Fast ion conductor, battery applications
Al_2O_3-SiO_2 (+R_2O, MO)		Structural refractories, firebrick Structural clay products Stoneware Porcelains Electrical insulation Chemically resistant coatings, enamels
SiO_2	diatom, zeolite	Lightweight refractory forms Ion-exchange media
Zirconia	ZrO_2, doped	Fast ion conductors (fuel cell and battery applications) Gas sensors Refractory thermal insulations Wear-resistant cutting tools
Magnesia	MgO	Electrical insulation Structural refractory forms
Titanate solid solutions		Piezoelectric devices (Pb, Zr) Microwave dielectrics (Zr, Ba) Optical memory (Pb, Zr) Ferroelectrics (Ba, Sr)
TiO_2		Pigments Humidity sensor (film)
Ferrites		Soft ferrites (Zn-Mn) Hard ferrites (Ba, Sr, mangetoplumbites, hexagonal ferrites)
Carbides, nitrides, oxynitrides	Al, Si, B	Cutting tools, abrasives Engine components Metal molds Metal filters Ceramic brazing compounds (complex formulas)
Phosphates, hydrates	Ca, Al, Si, Zn	Refractory castables Biocompatible cements Structural cements
Glass and glass ceramics		Consumer products Optical devices Fast ion conductors (advanced Li-batteries) Machinable ceramic forms Sealing glasses, glass ceramics

energy. The metallic, or electropositive, species are always separated by electronegative and highly polarizable oxygen. Separated charges define electronic dipoles, which will have a tendency to interact with an external electric field.

Ceramics containing transition metal cations may be paramagnetic or ferrimagnetic, interacting with magnetic fields. Electronic and ionic conduction may also be present, either as intrinsic or defect-based behavior. Electronic carriers can also interact with magnetic fields through precession.

A ceramic material interacts on an atomic level to any imposed electromagnetic field, electrical potential gradient, or electrochemical potential gradient. This interaction is the basis for devices that make use of the net polarization and magnetic moments of the lattice in the creation of unique devices that have revolutionized the electronics and communications industries. For device applications, the properties are usually optimized through phase equilibrium studies. It is found, for example, that the dielectric properties of complex titanate-based solid solutions are optimal at the stoichiometric or minimum defect composition. For all oxides, the dielectric properties of the lattice and its defects are responsible for continuing atomic scale processes, even at low and moderate temperatures.

Dielectric and Ferroelectric Interactions

The ability or capacity to store energy through dipole rotation or alignment is the material capacitance. The material dielectric constant expresses the effectiveness of the material versus free space in storing charge as defined using reference to a plate capacitor of area A and thickness d. If vacuum fills thickness d, the capacitance $C_o = (A/d)e_o$, where e_o is the permittivity of free space. For a material inserted into the gap, the capacitance $C = C_o(e'/e_o) = C_oK'$, where e' and K' are the relative permittivity and relative dielectric constant, respectively.

In most crystalline oxide structures, the net polarization of the individual dipoles is perfectly balanced to zero in the absence of an external field; these are dielectric ceramics. A net polarization represents a potential energy that can be stored, as in a capacitor, or converted to an alternate form of energy, as in a transducer.

A nonzero-sum lattice polarization can exist in crystalline structures that lack a center of symmetry. The periodicity of these low-symmetry lattices does not require an inversion or reflection of the dipole unit, which would have the effect of canceling the unit polarization. The noncentrosymmetric crystal classes are termed piezoelectric owing to their capacity to develop a static polarization or charge under stress. The dielectric constants and piezoelectric effects are directional in the lattice and are described through the tensor coefficients. The transition from piezoelectric structure type to dielectric is a first-order phase transition.

For all piezoelectric classes, there are strong and directional polarization interactions between electric fields, elastic and bulk strains, and thermal stress which are expressed through the coupled coefficients for electrostrictive, optoelectric, mechanooptical, and pyroelectric effects. First- and second-order phase transition temperatures and the characteristics of any of the coupled "effects" are readily altered or designed through phase chemistry and microstructural manipulation and have been the impetus for many phase diagram studies. Multicomponent oxides that crystallize in piezoelectric or ferroic structure types (see below) are good candidates for optimization through materials design and have largely replaced organic and hydrate crystal phases, which were among the original practical piezoelectrics.[1]

Ferroelectrics are a subclass of piezoelectrics in which the net polarization remains even after all stress or external fields are removed. The similarity between their characteristic polarization "loop" and the B×H loop of a square-loop ferrite, as diagrammed in Figure 2.1, is responsible for the designation "ferroelectric" and in no way relates to the properties of any iron-containing compound.

The "ferroic" structures are uniquely related to higher symmetry crystal classes (prototypes) through simple distortions. If two or more distortions of the higher symmetry unit cell exist that have the same internal energy and these variants are not related through a simple lattice translation, it is possible for oriented ferroelectric domains to exist in the crystal. The domains grow in response to external electric fields so that it is possible to "pole," or fully orient, the ferroelectric. The ferroelectric dipole ordering efficiency is temperature dependent and exhibits critical temperature behavior, similar to that exhibited in ferro- and ferrimagnetics' transition to the paramagnetic state. Thermal order/disorder transition to the paraelectric state is a second-order phase transition that may or may not coincide in temperature with a first-order, or structural, transition of the crystal lattice to a nonpiezoelectric class.

The tetragonally distorted perovskite structure in Figure 2.2 is an example of a practical ferroic type where a relatively small and highly charged *B*-site cation such as Ti^{4+} can interact with lattice O^{2-} and also the polarizable *A*-site cation such as Pb^{2+} or Ba^{2+} to form a permanent dipole. Near the critical or Curie temperature, T_C, the dielectric constants obtainable are extremely large.

Highly polarizable cations and spontaneously polarized electronic dipoles in a transparent ceramic or a glass can also interact with transmitted light to form electrooptical materials, where the polarization effect can be influenced by external applied electrical fields or mechanical strain.

The range of systems of interest for electronic application is large and growing. Some systems for electrical and sensing device applications are summarized in Table 2.2. Certain characteristics are listed for each application; however, secondary criteria such as strength, chemical resistance, or specific mass are often of equivalent importance.

It is important to remember that the complex field interactions that result in the practical ferroelectrics are to some extent active in all real ceramic materials.

---

[1]In these older materials, a low point-group symmetry molecule, such as OH^- ($_O/^H$) or H_2O ($_H/^O\backslash_H$), occupied a lattice point and could orient by rotation into the field. These were characteristically low-temperature materials with limited design range.

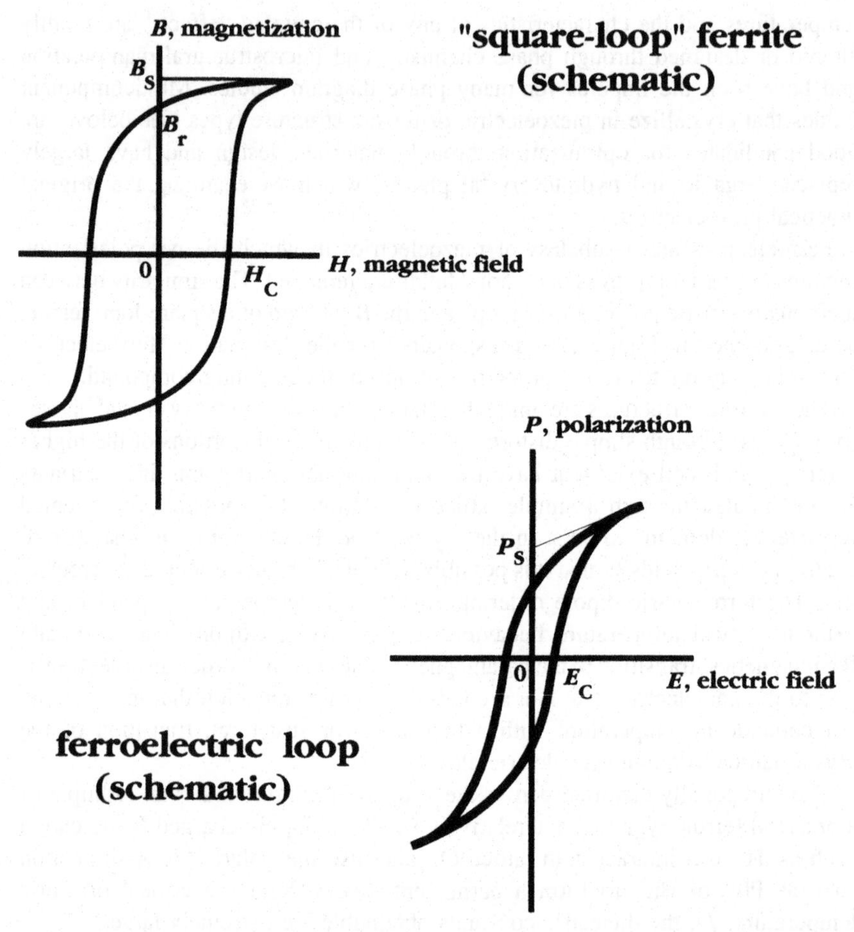

Figure 2.1 Characteristic square loop behavior in ferromagnetics and analogous ferroelectrics.

All real ceramics contain regions of reduced local symmetry owing to microstructural, surface, and impurity incorporation effects. Local dipole interaction can result in "lossiness" in any application subject to electromagnetic fields. Microscale strain and local heating effects may contribute to the rate of diffusive processes and reaction. Excessive "loss," or the inefficient conversion or transmission of electromagnetic energy, can lead to enormous thermal gradients and even localized melting.

Magnetic Interactions

Electric dipoles arise from separation of static charges inherent to ionic bonding, their relative ability to align with an imposed field results in dielectric behavior.

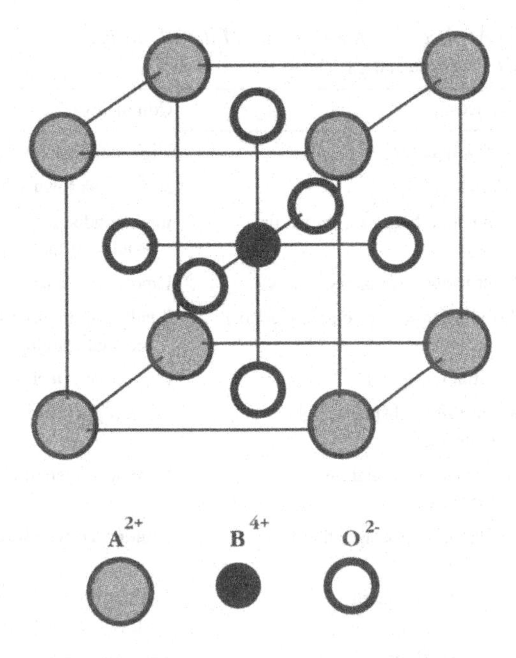

Ideal Perovskite ABO_3 Structure

Figure 2.2 Idealized perovskite structure, ABO_3.

The cyclical motion of electrons with unpaired spins, characteristic of the transition metals, similarly results in the formation of magnetic dipoles. Materials having inherently parallel aligned magnetic dipoles are said to be ferromagnetic; inherently antiparallel alignment cancels the net magnetic moment and antiferromagnetic behavior results.

True antiferromagnetism is rare; more likely it is the condition in which the alignment is not perfectly antiparallel with a resultant net nonzero permanent magnetic moment, a ferrimagnetic material. This is the most important class of ceramic magnetic materials.

Permanent magnetic moments are intrinsic to the atomic arrangements of oxide crystal structures which include transition metal species with imbalanced "*d*-level" electronic states such as Fe, Cr, Mn, Ni, and Co and also some lanthanide elements, particularly Sm. In the statistically randomized polycrystalline structures of conventionally processed ceramics, the overall magnetic moments are also randomized but can be aligned through the application of external fields. Crystallites of permanent magnetic materials can also be incorporated into polymeric composites to create recording media and flexible magnet tapes.

Materials with no free or unpaired electrons are diamagnetic, that is, their extremely small magnetic susceptibility corresponds to the slight ability of the

Table 2.2 Common Applications for Ceramic Materials in Electronic Devices with Significant Materials Characteristics

Material	Devices	Comments
Dielectrics	Disk capacitor	High stability, low cost
	Insulators	High breakdown voltage applications
	Solid-state batteries, fuel cells	Ionic conduction in electrically insulating material
	Microwave resonators, filters	Dipole resonance
Ferroelectrics	MLCC (multilayer chip capacitor)	Small size, high performance (with reduced stability)
	Oscillators, filters	Capacitive effects
	Light filters, displays, light deflectors	Electrooptic effect
	Thermistors, miniature thermostats	Nonlinear T-coefficient
	Holographic storage, memory devices	Photorefractive effect
Piezoelectrics	Frequency standards	Tuning resonance, radio, TV and other communications devices
	Selective filters	Resonance device
	Ultrasonic device, transducer	Resonance device
	Sonar	Resonance device
	Pressure guages, phonograph pickup	Nonresonant application
	Spark generators, ignition systems	Nonresonant application
Ferromagnatics	Permanent magnets	High H_C, B_S
	Magnetic bubble memory	
	Magnetooptical devices	
	Tape recording media	

bound electrons to precess under an imposed magnetic field. Materials containing magnetic dipoles that are inherently randomized, such as transition metal cations in solid solution, but that can be aligned under the influence of an imposed external field, are paramagnetic. The paramagnetic susceptibility decreases with temperature, as thermal fluctuations will tend to randomize the aligned dipoles. Diamagnetic susceptibility (which may be of similar magnitude) is not temperature dependent.

Possession of magnetostrictive properties, where the change in alignment of magnetic field results in a dimensional change, can allow the material to be used to convert electromagnetic potential to mechanical force, as in an acoustic device or driver. The reverse action would result in a pressure sensor.

Strong magnetic interaction is a relatively low-temperature phenomenon—thermal fluctuations tend to randomize either inherent or imposed ordering. The Curie temperature, T_C, represents the transition temperature for ferro- or ferrimagnetic to paramagnetic behavior. In the paramagnetic state, susceptibility decreases with temperature. The Neel temperature, T_N, is the corresponding transition temperature for antiferromagnetic to paramagnetic behavior. The transition to paramagnetic behavior will generally occur on heating well below $0.5T_m$, so strong magnetic interaction with the application environment will seldom affect properties or performance lifetime of a ceramic material in use at high temperatures.

In dielectric ceramics near room temperature, paramagnetic susceptibility arises from the presence of magnetic impurities, most commonly Fe, and is an important loss mechanism. Electromagnetic losses in oxide ceramics also result from the precession of free electrons that may be present from Fe and other transition metal impurities. Although strong interaction is a low-temperature phenomenon, free electron interactions with magnetic fields continue and may increase at high temperatures as ionization and electronic conduction increases. These processes contribute particularly to radiofrequency (RF) losses and heating effects.

Phase equilibria for the determination of magnetic transition boundaries in complex solid solution systems containing ferrimagnetic phases are an active field. Many important phases are based on related structure types: (1) spinel or cubic ferrite ($Fe_2O_3 \cdot MO$); (2) magnetoplumbite or hexagonal ferrite ($6Fe_2O_3 \cdot MO$); or (3) rare earth garnet ($3M^c{}_2O_3 \cdot 2Fe^a{}_2O_3 \cdot 3Fe^d{}_2O_3$) structures.[2]

Each of these structures has two or more cation sublattices of equivalent positions in which all magnetic spins are aligned and antiparallel to the alternate sublattice. (The superscripts in the garnet formula refer to the occupied sublattice.) Structure types and magnetic device applications of ferrite ceramic materials are summarized in Table 2.3.

Very low loss magnetic and dielectric materials can be used at microwave (1 to 100 GHz) frequencies. At these high frequencies, electrolytic and magnetic domain structures can no longer respond. Resonance, with retransmission or redirection of electromagnetic energy, results from material interaction.

Magnetic ceramics, particularly magnetic garnet-structured ferrites, are used in both microwave transmission (waveguide) and absorptive applications. The interaction of the ferrites' magnetic dipoles with the magnetic field of a permanent external magnet determines the precession frequency of the dipoles. Plane-polarized microwaves near that precessional frequency are absorbed and redirected, rotated by 45°. This gyromagnetic "faraday" rotation is the basis for devices to select, filter, and transmit microwave radiation. The capacity for microwave energy absorption is also used for shielding and "stealth" (radar absorbing) applications.

Microwave dielectric ceramic materials are high dielectric constant, very low

---

[2]Superscripts refer to the Wycoff designation of cation sites occupied in the rare earth garnet structure, space group *Ia3d*, no. 230.

Table 2.3 Ceramic Materials for Magnetic "Ferrite" Applications

Composition	Type	Properties/Applications
Spinel-based Ferrites: Soft "switchable" magnets		
$MO \cdot Fe_2O_3$ ($M = Fe,Zn,Ni,Co,Cu,Mg$) additives include CaO, TiO_2, V_2O_5, SnO_2, Al_2O_3 for property control	Mn–Zn ferrite	Low resistivity, high permeability, low-frequency applications
	Ni–Zn ferrite	High resistivity, moderate permeability, low losses to 10^6 Hz, some microwave applications
Magnetoplumbites, Hexagonal Ferrites: Hard "permanent" magnets		
$MO \cdot 6Fe_2O_3$ ($M = Ba,Sr,Pb$); ZnO, TiO_2 or Al_2O_3 substitution for high frequency applications	"Z" = $BaO \cdot 6Fe_2O_3$ "W" = $BaO \cdot 2MeO \cdot 8Fe_2O_3$ "Z" = $3BaO \cdot 2MeO \cdot 12Fe_2O_3$ "Y" = $2BaO \cdot 2MeO \cdot 6Fe_2O_3$	High anisotropy, high coercivity; used for resonance microwave devices at >50 GHz
Magnetic "Rare Earth" Garnets: Soft "switchable" magnets		
$3M_2O_3 \cdot 5Fe_2O_3$ ($M = Y$ or rare earth cation)		Low loss microwave applications, bubble memories

dielectric loss, solid solution compositions based on barium titanate, zirconium titanate, or barium tantalate. Typically, these phases exhibit a dielectric constant > 10 and inverse dielectric loss or $1/\tan\delta > 10^4$ at microwave frequencies. The resonance of the ceramic component is determined by the dimensions and dielectric constant of the ceramic material.

Effective microwave dielectrics exhibit very small temperature coefficients of dielectric constant and capacitance, which can be compositionally "tuned" to compensate for temperature drift due to other circuit components. Phase equilibrium studies in these materials are particularly useful in the determination of highly stable solid solution composition–temperature–pO_2 limits for reproducible processing results.

2.2 Refractory Applications

Refractory applications are at high temperatures, where the stiffness and relative stability typical of oxide crystal structures allow the material's use as containment for process heat and reactive species. The ceramic refractory is subject to extreme thermal, chemical, and mechanical stress and stress gradients. Additional interactions with the electromagnetic fields associated with power inputs to the process environment can be significant.

Excessive interaction with surroundings or contents, through conduction and transfer of heat or chemical species, will compromise performance and must be avoided; any reaction products must be noncontaminating. Refractories must also be stable in the process atmosphere, resist slump and creep at high temperatures for extended times, and be able to withstand large stresses during repeated heat-up and cool-down cycles.

A ceramic may also be subject to a hostile application environment without being an impervious containment. Examples are fibrous thermal insulations and reticulated or fibrous ceramic filters. Insulating microstructures—fibrous, bubble or low-density sintered products—must maintain their defined microstructural void spaces to continue satisfactory performance. In these cases, the strength and durability of the ceramic protects and maintains the functional internal structure.

Materials interaction in a chemically active application environment is considered corrosive when there is an electrochemical nature to the reaction, a transfer of charge or of charged species in a potential gradient. So long as the potential gradient exists and conductive paths are present, the interaction proceeds. Because of the nature of the charged species that compose the oxide lattice, solid-state reaction leading to compound formation or incorporation of solute from the refractory's environment is considered corrosive, as electrons and charged ionic species must be transferred between reactants and products.

"Passivation" occurs when the product of corrosive reaction is insulating and compact (uncracked and continuous). "Passivation layers" on metals are dense, coherent oxides that block further reaction. Similarly, multicomponent reaction products may form between oxides, carbides, and/or nitrides which block further mass transport between reactants. Limited surface reaction is the basis for extended survival in many refractory applications and in many varied ceramic applications not generally thought to be refractory, such as sealing glasses and glass ceramics or the complex metal carbides used as brazing compounds in advanced ceramic forming. Successful joining requires a graded reaction zone in which the reaction products have compatible mechanical and thermal properties.

Development of successful ceramic processes and products requires understanding of surface reactions, the conditions for stable and metastable melting, and the nature of possible product phases. The phase diagram specific to the local interaction system relates the initial chemistry and phase structure to the expected properties of a postulated reaction, corrosion or passivation product.

The development of even the simplest equilibrium-based model for the behavior of a system can enable the engineer to interpret and use the information available from production data, routine experiments in product or process development, or even from the most catastrophic failure. The early application of available phase diagram information is perhaps the most effective means to anticipate materials performance and prevent avoidable failures. The phase diagram and reaction equilibrium in the complex systems pertains directly to the performance of many refractory containments.

3

Mass Transport for Reaction

Oxide structures contain two or more sublattices in a prototypical array that defines the ideal dielectric, magnetic, and electronic structure. The requirements of symmetry and charge neutrality require that the creation of a defect, a change on one ionic or structural sublattice, be reflected in a balancing change on one or more of the other ionic or structural sublattices or through the creation of a charged electronic defect. Owing to the requirement of continuous charge balance and conservation of lattice symmetry and continuity, reaction and corrosive interaction is relatively slow in close-packed ionic lattices and relies on the preexistence of defects to progress.

3.1 Defects in Oxides

In the previous section, the ferroelectric or ferroic structures were described with reference to a higher symmetry prototype. Similarly, mass and charge transport is most readily described with reference to a perfectly insulating dielectric oxide prototype structure.

The periodicity of the ionic crystalline structure defines the normal lattice occupancy. Any deviation from the prototype then can be described as a defect with a contribution to the lattice energies and also an "effective charge."

The introduction of a defect species also defines a new state in the density of states. The density of states, together with the identity and relative mobility of charged species in the ceramic, determines whether the defective material will behave as a semiconductor, narrow band electronic conductor, ionic conductor, or exhibit mixed conduction. The conductivity "type" is determined both by the proportion of charge carried by each species (i.e., electronic or ionic carriers) and the temperature characteristics of the mobility. Undoped semiconductors and ionic conductors have a thermally activated mobility, whereas narrow band electronic conductors (in which the mobility may be of the same magnitude as the former in a particular temperature range) will show a linear decrease in mobility with temperature due to phonon scattering, an interaction with the vibrational modes of the lattice. The measured conductivity is the product of the carrier concentration, carrier charge and mobility, summed over all species present.

There are conductive oxides, generally the transition metal oxides, in which there is an equilibrium between the cation valence states. In the density of states, this equilibrium is reflected through an overlap of occupied and available *d* or *f*-type levels into a conduction band. Electrical conductivity in materials such as CrO_2 and ReO_3 can approach that of metals. Dense sintered electrically conductive ceramics will generally appear black (or very dark if powdered), whereas dense semiconductive ceramics exhibit a color that is characteristic of the discrete energy transitions of the electronic carriers, often in the range of visible light.[1] In electrically conductive ceramics, lattice defects reduce overall conductivity through increased phonon scattering. In semiconductive or insulating ceramics, the presence of lattice defects generally increases overall conductivity as the carrier concentrations rise (unless defect clustering decreases mobility).

Ionic conduction, the motion of charge via atomic defect species in an applied potential, is found in all real oxides that contain lattice defects at any temperature above 0 Kelvin (K). Practical ionic conductors have >90% of charge transported by ionic species and can be achieved in some materials, particularly "stabilized zirconia" and the beta-aluminas, through solid solution design for suppression of electronic carrier population (as discussed later in this chapter).

Ideal dielectric properties are defined by the crystalline order and atomic makeup, while semiconduction and ionic conductivity behaviors in most oxides largely result from disorder. This allows the materials designer great scope in property development and optimization through intentional variation of phase chemistry, defect structure, and microstructure. Special attention has been paid to phase equilibria of solid solution boundaries, as breakdown of the solid solution may decrease efficiency. However, the more general effects of impurities, lattice defects, and microstructural flaws on reaction rate kinetics due to increased charge and mass transport are essential to practical ceramic processing and long-term behaviors.

A defect is generally described relative to a perfect periodic structure or prototype. The ideal structure extends infinitely in space in three dimensions; thus any surface or interface is also a defect. Liquids also have a "thermodynamically ideal" structure, as do gases. Their prototype is perfectly random and without

---

[1]It is interesting to watch a ceramic as it cools from the furnace after sintering or heat treatment. After cooling below about 1000°C (black body radiation above that temperature makes the piece appear yellow-orange), many go through progressive color change, from dark to light on cooling, often with a distinct color or color change which is indicative of these electronic transitions.

chemical interaction between particles. A "defect" in a random structure can be defined as a local molar volume change related to atomic configuration or molecular conformation.

Random defects are thermodynamically favored in real condensed phases above 0 K, as they make a real contribution to the configurational entropy, reducing overall free energy. Any notionally perfect crystal at finite temperature still contains defects due to atomic fluctuations and uncertainty. The ideal "X-ray" crystallographically derived lattice is a time-averaged image of a real structure composed of particles that are in constant motion. The point and lattice symmetries which are assigned to the real structure are convenient mathematically and conceptually to describe the properties of the statistical ensemble composed of many hundred billion individual particles. The true atomic positions in any real structure are subject to great actual uncertainty.

In a reacting system, defects enable the growth of a new crystal lattice through destruction of the old lattice or lattices. A site balance of defects must also be maintained in the new and old lattices throughout a reaction or transformation process.

Defect Terminology

Intrinsic defects arise as a result of the constant vibrational motion of atoms at nonzero temperature and are always present in a real material. Defects arise also in response to mechanical constraints and lattice truncation at interfaces that result in local charge imbalance or nonstoichiometry. Extrinsic defects are created to balance the effects of dissolved impurities which also cause lattice strain and charge imbalance. Some of the more common defect species that occur are illustrated in Figure 3.1.

Atomic defects are of three basic types: an atom where no atom would normally reside—an interstitial; an unoccupied normal atomic site—a vacancy; or the wrong atomic species on a normal lattice site—a substitutional atom. Atomic substitution of one cation for another or one anion for another is common in compounds having two or more cation or anion types and is also a normal means of impurity incorporation. "Antisite" disorder, or cation–anion substitution, is highly unfavorable in ionic compounds for electrostatic (internal energy) considerations. Antisite disorder is found in covalent compounds and intermetallic compounds.

The defect's charge is defined *relative to* the perfect atomic lattice. A notation that is widely accepted for the description of defects in nonmetallic systems is attributed to Kröger. The symbol is of the type $(\text{Species})_{(\text{site})}^{(\text{charge})}$.

"*Species*" is the atomic or quasichemical species symbol, V = vacancy, e = electron or h = electron hole; "*charge*" is either the number of dots ($^{\bullet}$) for positive or primes ($'$) for negative (*relative*) charge, and "*site*" refers to the perfect crystal lattice site on which the defect is found and is identified by the normal lattice

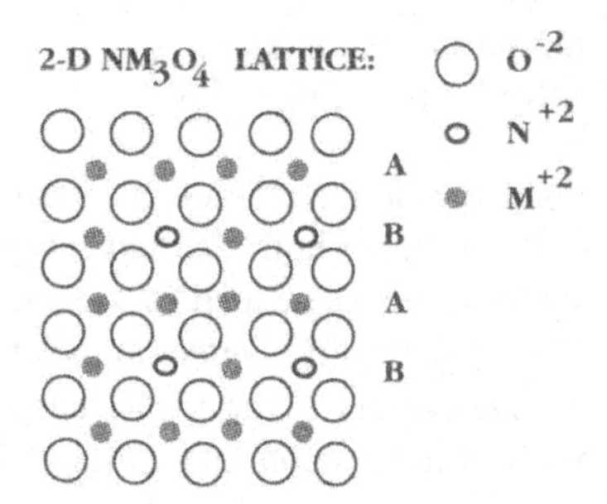

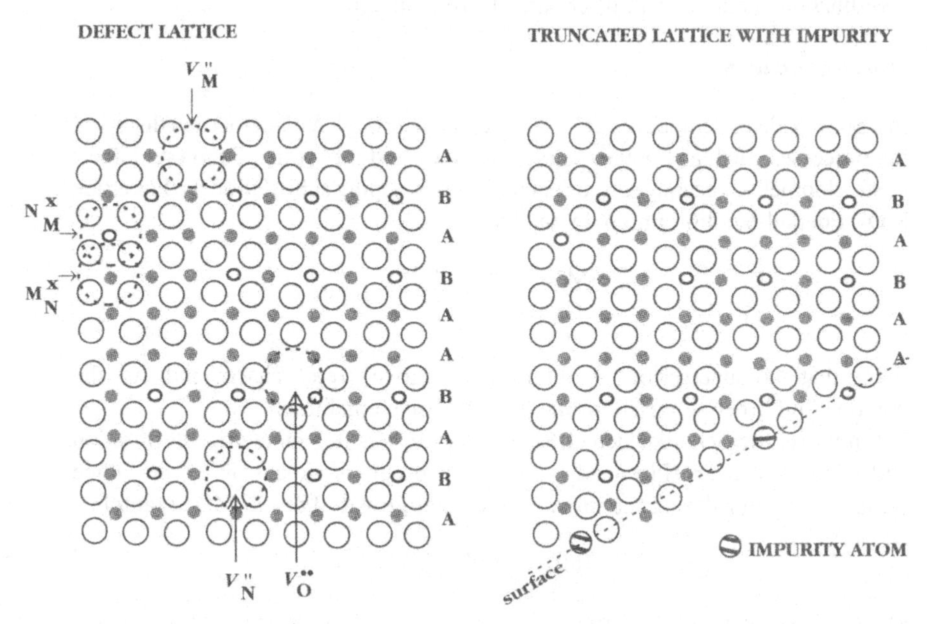

Figure 3.1 Illustration of defect species in the two-dimensional NM_3O_4 lattice. The upper drawing represents the perfect lattice. The normal stacking sequence is A–B–A–B. A stacking fault A–B–A–A–B–A is illustrated at lower left. Several vacancy and substitutional species are shown. (Interstitial species are not favorable in this close-packed lattice plane, although interstial sites exist above or below the plane of the paper.) At lower right, truncation of the lattice, as at a surface, with adsorbed impurity stabilization is shown.

species, by "i" for interstitial, or by the site type (such as A or B sites in a spinel structure).

Examples:

Al'_{Si} = Al^{3+} substitutional cation on a normal Si^{4+} cation site.
$Li_i^{•}$ = Li^{1+} cation on an interstitial (not normally occupied) position.
$Cl_O^{•}$ = Cl^{1-} substitutional anion on a normal O^{2-} anion site.

$V_O^{\bullet\bullet}$ = Vacancy of a normal O^{2-} anion site.
V_{Mg}'' = Vacancy of a normal Mg^{2+} cation site.

Cations, as substitutional or interstitial defects, will generally be found to occupy a coordination site that satisfies both size and charge requirements. Defects of opposite charge are likely to be associated in the local atomic environment, for the maintenance of charge neutrality. Defect association may extend beyond "first shell" coordination, particularly if the local charge density over first and second (plus perhaps the third next anion and cation species) nearest neighbor coordination shells is neutral, or sums to zero charge.

Defect Formation

In a pure oxide or compound, defects are created through natural thermal disordering processes, such as Frenkel pairs of vacancy and interstitial, Schottky defects of stoichiometric vacancy sets, or pairs of site-swapping substitutionals. For oxide MO_2, Frenkel equilibrium could be written:

$$M_M^x \Rightarrow V_M'''' + M_i^{\bullet\bullet\bullet\bullet}$$
$$O_O^x \Rightarrow V_O^{\bullet\bullet} + O_i''$$

[Note that the superscript "x" refers to a relative "zero" charge and that the notation "nil" (below) refers to the perfect lattice condition.]

Similarly, the Schottky defect for this lattice is a stoichiometric set of anion and cation vacancies. Physically, this might occur on expansion of the lattice during heating or during the process of crystal growth. For the MO_2 example

$$(\text{nil}) \Rightarrow V_M'''' + 2V_O^{\bullet\bullet}$$

illustrates the Schottky equilibrium. For a compound with two cation species, such as $A^{2+}B_2^{3+}O_4$, Frenkel disorder could exist on either A or B cation sublattices (each would have a unique equilibrium constant and energy of defect formation) and Schottky disorder would be written

$$(\text{nil}) \Rightarrow V_A'' + 2V_B''' + 4V_O^{\bullet\bullet}$$

Site substitution at equilibrium is described

$$A_A^x + B_B^x \Rightarrow B_A^{\bullet} + A_B'$$

Compound stoichiometry is always maintained in the formal description of intrinsic defect equilibrium. In a reacting system, where defects enable the growth of a new crystal lattice through destruction of the old lattice or lattices, a site balance of defects must also be maintained in the new and old lattices. For example, if a new cation is added to the growing MO_2 lattice, two oxygen sites (currently vacant) are also added to maintain the site balance and stoichiometry. The surface acts as both a source and sink for atomic lattice defects and may act similarly for electronic defects through interaction with the environment. The presence of surfaces is necessary for defect processes to proceed in real ionic systems.

Electronic defects similarly describe the presence or absence of electrons in the electronic structure, or normal density of states. An excess electron, relative to the normal requirement for bonding in the perfect lattice, is a carrier of negative charge which may be more or less free to move in the structure. Its energy is higher than that of a bonding electron. If the energy is sufficient, it may occupy a previously empty conduction band state. A lower energy electron may reside in new "defect state." A missing electron is termed an electron hole, and is an empty state in the valence (or bonding) electronic energy levels. It can act as a positive charge carrier or as a "sink" to consume excess electrons from higher energy states, reducing the net energy of the structure.

Electronic defect pairs are created thermally or by dissociation of charges from ionic defect species, but are governed by the equation

$$(\text{nil}) \Rightarrow e' + h^{\bullet}$$

which is described by the quasichemical equilibrium

$$pn = k_i = \exp\left(\frac{\Delta E_{\text{gap}}}{k_b T}\right)$$

$$p = [h^{\bullet}],$$
$$n = [e']$$

It is interesting to note that defect formation energies are often stated as "per defect" and not for the equilibrium stoichiometric relationship when discussed in the literature. Stating "per defect" formation energy is a useful convention when comparing the formation energy of a cation vacancy in its pure oxide with that in its compounds, but can be confusing if not clearly identified. Experimental defect formation energies are actually formation enthalpies and are usually found in the range of 1 to 5 *ev per defect.*

Extrinsic Defects

Real materials are not pure and contain at least several ppb (parts per billion) chemical impurities. Materials containing only 10 ppm (parts per million) are considered reagent grade in general discussion. In perspective, 10 ppb corresponds to about 10^{15} per cm^3, which is usually on the order of the intrinsic defect population at 1500°C. In real materials the defect and transport properties that allow reaction to proceed in the solid state at almost all temperatures of interest are governed extrinsically by impurities.

The presence of the impurity does not in itself greatly change the fundamental energetics of reaction. Rather, the compensatory defects formed to balance the charge or relieve strain will reduce the additional energy needed to further disrupt the "perfect lattice" so that an atom might change position, greatly reducing the activation energy of reaction. Charged defect and impurity species association can also result in lattice dipole formation. Thus, impurities can interact with electromagnetic fields, resulting in localized "lossy" behavior and thermal effects that can enhance nucleation or reaction rates.

An impurity species may be incorporated in the host lattice in a number of ways. The most likely will be that which results in the least lattice strain and smallest local charge imbalances. Substitutional solid solutions have high stability for cationic species of the same formal charge and similar size ($\leq$ 15% difference in size).

Defect equilibria are discussed in the dilute solution regime, with noninteracting defects. For extrinsic defects, the dilute regime ends when there is a significant probability that an impurity ion will "see" another impurity ion on the adjacent (or next nearest neighbor) site—generally at about 0.1 cation equivalent percent (1000 cation ppm) from simple statistical considerations. Effective shielding by the oxygen sublattice may extend application of the dilute solution models to several percent.

At higher concentrations, clusters will begin to form that may appear very like the basic building blocks of probable compounds in the system of the base plus impurity oxides. Clustering becomes more probable for cation solutes of very different electronegativities or ionization potential, compared either between various solutes or to the host. Compound formation is favored over solid solution in such cases.

If a number of impurities are present, it is the total concentration of similar species[2] that will determine whether the dilute solution approximation is appropriate. Extension of the dilute solution model to high defect concentrations (>5 to 10%) is routinely applied for simplistic interpretation of transport data. This is a reasonable first approximation but will probably fail in detailed examination of transport or crystal chemical data.

The formalism of extrinsic defect equilibrium is stated using quasichemical reactions to describe the related changes on each sublattice resulting from an

[2]Similar species are similar in charge, size, and electronegativity; generally members of the same family of the periodic table may be grouped together. Extrinsic defects may also be compensating; for instance, the net effect of the 1+ cation species may cancel all or part of the effect of the 3+ cation species in an AO-type oxide, with the result being a low effective defect concentration in a very impure material. While the defect-related transport kinetics might then be characteristic of a fairly pure material in that case, the phase equilibria would reflect the total impurity concentration.

incremental chemical change. The defects that compensate the impurity must be in equilibrium concentrations as defined by the Frenkel, Schottky, and electronic defect formation relationships of the pure material. Overall charge and site balance, electroneutrality, and stoichiometry must be maintained.

3.2 Ideal Impurity Incorporation

As an example, consider MgO as a dilute impurity in Al_2O_3. Limited solution is expected as the difference in ionic radius is over 20% in six-fold coordination[3] as typifies corundum. The quasichemical reaction for substitutional incorporation is written

$$2\text{MgO (in Al}_2\text{O}_3) \Rightarrow 2\text{Mg}'_{\text{Al}} + 2\text{O}^{x}_{\text{O}} + V^{\bullet\bullet}_{\text{O}}$$

(The oxygen vacancy that forms maintains site stoichiometry, as the lattice is extended to allow incorporation of two MgO molecules.)

In Al_2O_3, Schottky equilibrium is stated: $(\text{nil}) = 2V'''_{\text{Al}} + 3V^{\bullet\bullet}_{\text{O}}$
Anion Frenkel equilibria: $(\text{nil}) = \text{O}''_{\text{i}} + V^{\bullet\bullet}_{\text{O}}$
Cation Frenkel equilibria: $(\text{nil}) = \text{Al}^{\bullet\bullet\bullet}_{\text{i}} + V'''_{\text{Al}}$

The ionization of the electronically charged species must also be described through equilibrium relationships such as $V^{\bullet\bullet}_{\text{O}} + 2e' = V^{x}_{\text{O}}$ and $V'''_{\text{Al}} + 3h^{\bullet} = V^{x}_{\text{Al}}$ (these are generally multiple-step reactions).

The total charge balance for electroneutrality in the system will be stated as a summation of all positive versus all negative charged species:

$$3[V'''_{\text{Al}}] + 2[\text{O}''_{\text{i}}] + [\text{Mg}'_{\text{Al}}] + n = 3[\text{Al}^{\bullet\bullet\bullet}_{\text{i}}] + 2[V^{\bullet\bullet}_{\text{O}}] + 2[\text{Mg}^{\bullet\bullet}_{\text{i}}] + p$$

Realistically, defects having relatively small charge and small strain are most favorable, eliminating the triply charged species from normal consideration. Interstitial species are usually less favorable than simple vacancies or substitutional defects in a close-packed lattice, although the corundum structure has normally vacant anion polyhedra that could be partially occupied by $\text{Mg}^{\bullet\bullet}_{\text{i}}$, perhaps in association with an oxygen vacancy or other relative negative defect to satisfy local charge balance. Consideration of the close-packed nature of the oxygen sublattice leads to elimination of the O''_{i} as a likely species owing to electrostatic and lattice strain. Electroneutrality for dilute (unassociated) defect concentrations could then be simplified to the most likely, or lowest charge and lowest strain, individual species:

$$[\text{Mg}'_{\text{Al}}] + n = 2[V^{\bullet\bullet}_{\text{O}}] + p$$

[3] See Shannon and Prewitt (1976), "Revised effective ionic radii."

In alumina and other wide band gap insulators, the electronic species would not be considered important species for electroneutrality until very high temperatures.

Further consideration of physical processes of defect formation and lattice relaxation at various conditions of temperature and pressure leads to the recognition of several simplified neutrality regimes, and allows "stepwise" consideration of the effect of temperature and chemical potential on the populations of mobile species required for reaction to proceed.

If multiple solutes are present, the net effect may be the creation of offsetting defects. For instance, introduction of SiO_2 into Al_2O_3 would be written

$$3SiO_2 \text{ (in } Al_2O_3) = 3Si_{Al}^{\bullet} + 6O_O + V_{Al}'''$$

(Incorporation of extra oxygen as interstitials is not considered due to excessive strain.)

The incorporation reactions for MgO and SiO_2 in Al_2O_3 can be combined and result in a recombination of compensatory Schottky pairs of defects. A material with balanced cation substitutions could behave as a stoichiometric oxide in lattice transport studies, having minimal lattice vacancies and thus a low diffusivity, but would be totally unsuitable for low-loss application in electromagnetic fields.

Clustering of impurities defects can be beneficial to the dielectric loss and other transport-related properties. A low strain cluster or precipitate, possibly of a stoichiometry related to cordierite ($2MgO \cdot 2Al_2O_3 \cdot 5SiO_2$), could be formed through appropriate heat treatment as determined using the phase diagram. The effectiveness of "sintering aids" (oxide dopants) in improving the properties of dielectric ceramics is often the result of their tendency to form a precipitate phase that incorporates (or "getters") solute species that would otherwise degrade the oxide's desirable technological properties. A pleasant side effect of this type of chemical design can be an improved microstructure, if the precipitate phase can be engineered to act either to nucleate new grain growth or to pin existing grain boundaries. (See discussion of sintering aids in Chapter 8.)

3.3 Impurities and Defects in Real Oxides

Very large concentrations of point defects are not found in large insulating oxide crystals; the increased strain and internal energy rapidly outweigh the favorable entropy contribution. In the ionic structure, stable point defect populations require the presence of an unconstrained or free surface to act as a source or sink for vacancy and substitutional species and also to allow for relief of the lattice strain that accompanies interstitial or Frenkel defect formation.[4] The surface itself is

---

[4]In metals and some covalent crystals, metastably large point defect concentrations can coalesce to form line and planar defects (dislocations and shear planes). Stability is maintained during crystal growth, reaction, and plastic deformation of the crystal as these extended defects can act as both source and sink for point defects.

associated with an electrochemical potential gradient affecting defect populations within its characteristic dimension, the debye length. The debye length is defined

$$\partial = \left(\frac{\varepsilon\varepsilon_o k_b T}{e^2 \Sigma_i (N_i z_i^2)}\right)$$

where ε = material dielectric constant
ε_o = permittivity of free space
e = electron charge
N_i = number density of species "i" having relative charge z_i

For 100 ppb, N_i is on the order of 10^{21} cm^{-3}. The debye length is thus on the order of several 10s of nanometers, for even a very pure material with reasonably clean surfaces.

The debye length corresponds to a reasonable fraction of the crystal lattice that is highly defective. The structures of the near-surface regions of oxide particles are not simply defined by the stoichiometric requirements of the bulk phase. Rather, the structures will reflect the stoichiometric requirements of the local chemistry which include high solute concentrations and adsorbed species.

For example, the near surface structure of Al_2O_3 containing segregated Ca^{2+} has been shown to be very like $CaAl_{12}O_{19}$ ($CaO{\cdot}6Al_2O_3$), requiring ordering over several cation layers. The phase diagram for the CaO–Al_2O_3 system indicates that this stoichiometry is compatible with Al_2O_3.

Many ferroelectric titanate phases show varied stoichiometry in the near-surface region, also indicating the effects of surface electrochemical gradients. Within the debye length, 10 to 15 nm being commonly noted, major cation stoichiometry is altered with accompanying solute depletion or segregation. In very fine particle sized powders, the near-surface effects can be sufficient to "stabilize" a nonequilibrium phase or to otherwise effect phase transition behavior. Complex ordering in near-surface domains can also be stabilized by interaction with surface adsorbed species, particularly protons which can percolate along the domain boundaries.

The formation of ordered defect structures that are similar to the thermodynamically compatible phases of the phase diagram is a predictable response of a segregated, impure chemical system. Near a free surface, defect ordering can proceed under minimal volume constraint, minimizing or eliminating lattice strain without resulting in discontinuities in chemical potential gradients. Ordering into compatible structures also minimizes the number of charged defect species present in the near surface, reducing the slope of the electrochemical potential gradient.

The phase diagram is applicable in the most common situations involving nonvolatile solutes. In the example of Ca segregated in Al_2O_3, the phase diagram applies as the solute is not mobile or volatile—the system is closed. Near-surface defect types and their organization and structure can be hypothesized from size (strain) and charge information on solutes and available surface species. Atomistic simulation techniques can be effective in the generation of plausible near-surface equilibrium structures from compositional and crystal chemical data; however, the phase diagram is an important first reference for a starting point even for advanced calculation methods.

For proton stabilization of domain structure, the system is open to interaction with surface volatile species such as water vapor. The mobile species, H^+, must have a known constant activity, such as from solution pH, for a simple relationship between component activity and structure to be established. (See discussion of mobile components in Chapter 6.)

More generally, the mobile species is oxygen. Systems with variable valent cations have characteristically broad solid solutions with a large tolerance for solutes of varied size and charge and a range of oxygen stoichiometry. For interpretation of near-surface effects on phase transition behavior, the local stoichiometry must first be determined as there is no exact correlation with bulk composition. The phase diagram can serve as a guide to interpretation of segregation and surface effects on a phase of a known, fixed stoichiometry. The phase diagram of a system in which a species is volatile, with structure or stoichiometry seen to vary with its activity, must be presented at constant activity of that species for the diagram to be an effective tool in process design or for property interpretation.

4

From Equilibrium Thermodynamics to Phase Equilibria

Traditional thermodynamics was developed prior to the understanding and broad acceptance of the atomistic nature of matter and the crystalline state. The fundamental equations of equilibrium thermodynamics are equally suitable in the description of macroscopic or microscopic phenomena of energy transfer and balance in mechanical or chemical systems. However, meaningful application of thermodynamic quantities and relationships in understanding phase equilibrium cannot be accomplished without full awareness of the nature of the phases under consideration.

At equilibrium, the observed state of any system represents a balance between the enthalpy (H) and entropy (S) of the system:

$$\Delta G = 0 = \Delta H - T\Delta S$$

In a thermomechanical system of traditional thermodynamics, the enthalpy change (ΔH) is the maximum heat energy that must be supplied (endothermic process) or could be extracted (exothermic process) in "reversibly" realizing the equilibrium state from an initial state, while the entropy change (ΔS) is that amount of heat energy that is "lost" (unavailable for performing useful work) owing to permanent changes that have occurred in the system in realizing the final state.[1] The internal energy (E) is related to the enthalpy through the equilibrium relationship $\Delta H = \Delta E + P\Delta V$ [Alternatively, $\Delta E = q - w =$ (heat) $-$ (P–V work)]. The definitions of these quantities and their manipulation in describing thermodynamic changes of state of a macroscopic thermomechanical system are found in any standard text on the first and second laws of thermodynamics and will not be detailed here.

---

[1]Entropy change, defined as q/T for either system or surroundings, is best determined from the sum of entropy changes found in a reversible cycle.

Thermodynamic quantities assess the potential for a physical system to undergo a change due to its environment or to effect a change on its environment. If, on the addition of a small increment of energy or mass from its environment, the system can achieve a new configuration of lower energy, the original system is unstable. Equally, if the system can release a small increment of energy or mass to its environment and achieve a configuration of lower energy, the original system is unstable. The configuration having no potential for change to a lower energy state is called "equilibrium," and is defined for specific, constant conditions of temperature and pressure. Fine-scale heterogeneity and chemistry are not considered in classical thermodynamics.

Gibbs modified the definition of "system" such that its heterogeneous nature could be considered. In Gibbs' system, mass is conserved in the system as a whole, but within the system are phases having uniform properties whose proportions reflect an equilibrium distribution of components. The composition and number and proportion of phases present are uniquely defined by the state variables, which include the system composition, and will be unchanging at equilibrium. The Gibbs system definition and the "phase rule" are discussed more completely in Chapter 5.

4.1 Thermodynamic Functions and Relationships

In unconstrained condensed or vapor systems, the most usual statement of the condition of the system is the Gibbs free energy, "G" or "ΔG". Discussion based on the Gibbs free energy is most useful in constant-pressure systems. The Helmholtz free energy (defined as $\Delta A = \Delta E - T\Delta S$, where $\Delta H = \Delta E + P\Delta V$ and E is the internal energy) is most useful in describing the equilibrium condition of constant volume systems. An example of a constant volume system in ceramics processing would be reaction and equilibration of crystals constrained in a dense crystalline or other nondeformable matrix. Reaction in a "bomb" calorimeter, commonly used to investigate high-temperature and high-pressure equilibria, are also generally considered as at constant volume. All discussions of equilibrium in terms of either the Gibbs or Helmholtz free energies are consistent, but these are not equivalent quantities. In this discussion, the "free energy" refers to the Gibbs free energy, unless otherwise clearly stated.

At equilibrium, $\Delta G = 0$. The Gibbs function, or free energy of formation of a condensed phase, is often expressed in terms of the enthalpy and entropy of the system: $\Delta G = \Delta H - T\Delta S$. In a condensed crystalline system the statement quantifies the competing tendencies of electrostatic and "crystal field" or Madelung energy (i.e., bonding and order) versus entropy (or thermal energy and disorder) in the determination of the equilibrium state of the system. At equilibrium, a balance is achieved and the structure is stable.

The "delta" (Δ) notation implies *either* that all the energies are defined with reference to a known stable state *or* that all energies are defined with reference to a convenient and definable state. The equation refers to a change in state from initial or reference to final. Classical thermodynamics allows for any path between states to have the same net energy quantities and, as both initial and final state energies are described with reference to a known state, the actual reference state is technically inconsequential. This technicality enables elegant derivations for equations of state for highly nonequilibrium systems, providing a common reference state (even one that cannot be physically realized) can be defined. An example of such a notional reference state in phase equilibria would be a mechanical mixture of pure liquids as the reference state for liquid or solid solution formation.

Similarly, a thought experiment to calculate the free energy of formation of alternate condensed phases formed from their elemental standard states can be performed by means of application of the Born–Haber cycle. Atoms and their electrons are theoretically separated in vacuum to their ideal ground states and then reassembled in a crystal lattice of known symmetry. Each electron in each bond is separated from the ground state atom to infinity and then transferred to the bonding state in stepwise fashion, utilizing tabulated values for ionization potential and electron affinity. Each infinite lattice of point charges (ions) is characteristically defined as having a Madelung energy which relates to the internal energy of the resultant crystalline structure. No single step in this process can be physically achieved, yet each reference state is fully defined.

The means of theoretical calculation for lattice energy, compound formation and defect formation are discussed in primary references on the subject. These methods and their successful applications illustrate the importance of the atomistic nature of matter in formulating a thermodynamic model for materials.

Multicomponent Equilibria and the Chemical Potential

In a multicomponent system at constant temperature and pressure, the change in Gibbs free energy of an individual phase due to an incremental change in the concentration of one component is defined as the chemical potential.

$$\mu_i = \left(\frac{\partial G}{\partial n_i}\right)_{T,P,n_j}$$

Nonuniform composition or metastable phase development results in local energy gradients which are the driving forces behind the continuous trend of atomic movements toward the equilibrium state.

The chemical potential is readily included in a general differential statement of the Gibbs function of each phase. The base equation is simply derived from the definition of the Gibbs free energy, where $G = H - TS$ and $H = E + PV$. Differentiating both equations, the result can be combined for equilibrium as

$\partial G = V\partial P - S\partial T$ in the absence of a chemical contribution. For a multicomponent system, the additional differential term for the chemical potential is included:

$$\partial G = V\partial P - S\partial T + \Sigma\mu_i\partial n_i$$

where the final term is summed over all species.

The differential form of the Gibbs equation is the basis for all thermodynamic relationships linking the observable properties, such as molar volume and concentration, and experimental intensive variables (T and P) to the energy of the system. Under conditions of constant temperature and constant pressure, the relationship simplifies to : $\partial G_{T,P,n_j} = \Sigma\mu_i\partial n_i = 0$ at equilibrium.

A finite and nonzero Gibbs free energy is also interpreted as a measure of the capability of the system to perform useful work. In thermodynamics, work is either the mechanical (springs, pendulum, etc.), P–V type (steam or carnot cycle) or can be associated with a change in composition at constant temperature and pressure. The potential to do work as a result of compositional change is the chemical or electrochemical potential of the system and has no mechanical analog.

The term "chemical potential" is easily understood in terms of a typical battery. The voltage (or electrical potential difference) arises because of the difference in concentration of a common ion between the separate chambers, and is measurable because of the electron motion (which pass more easily through a voltmeter or an external circuit) necessary for a change in bonding of that common ion. The measured voltage can be shown to be exactly equal to the chemical potential difference, and the electron current can be used to do work proportional to that potential difference and the mass of material available in the battery. The Gibbs free energy of a postulated reaction can often be equated to or determined from tabulated electrochemical half-cell potentials if the reaction in question can be described in appropriate terms involving electron transfer-type reactions.

Partial Molar Quantities

In multicomponent systems where phases can exist within a range of compositions (liquids, glasses, or crystalline solid solutions), it is often convenient to refer to "partial molar" quantities. At fixed conditions of temperature and pressure, the free energy balance can be stated using the Gibbs equation expanded here for the components a, b, c. . . .

$$dG = \sum_i \mu_i\partial_i = \mu_a\partial_a + \mu_b\partial n_b + \mu_c\partial n_c + \ldots$$

Each of the m_i are the chemical potentials of the component "i," or "partial molar free energies," and δn_i is the incremental change in composition of the phase in terms of that component. Partial molar quantities are usually determined by the "method of intercepts," by taking the tangent of the physical property curve as described in Figure 4.1, and are meaningful only over the range of observable composition of a particular phase, which may be very narrow (a so-called line compound) or extensive (liquid or solid solutions).

The experimental determination of partial molar quantities is difficult at best. They are most readily determined through the use of electrochemical concentration cells or measurements of equilibrium partial pressures of components over the condensed phases of interest. Calorimetry may be used to determine partial molar heats of solution in some systems. The determination of partial molar volume and its temperature dependance is readily performed using dilatometry, while determination of composition–volume relations in quenched systems can be performed with great precision using lattice parameter determination. These latter measurements, not performed on equilibrium specimens, are acceptable for the qualitative comparison of similarly prepared specimens only.

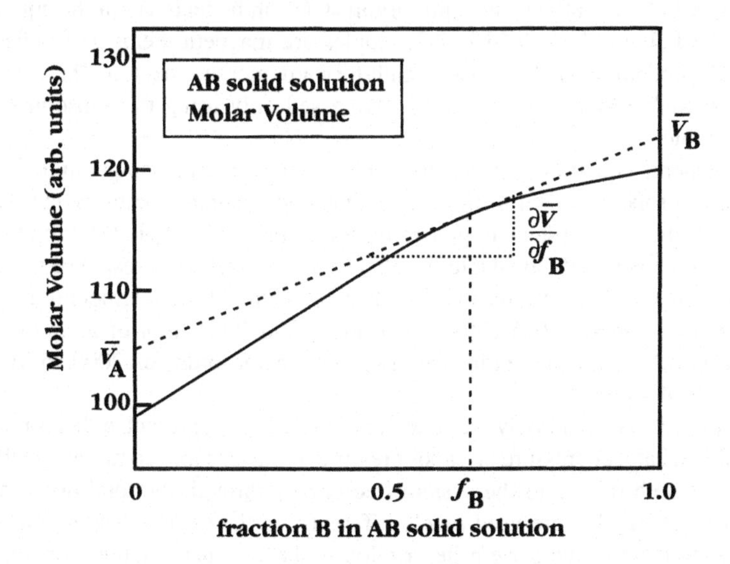

$$\text{Solution Molar volume} = \bar{V}_A + \left[\frac{\partial \bar{V}}{\partial f_B}\right] f_B$$

$$= \bar{V}_A (1 - f_B) + \bar{V}_B f_B$$

$\bar{V}_A$, $\bar{V}_B$ are partial molar volumes of components A and B in the solution.

Figure 4.1 The method of intercepts demonstrated for the determination of molar volume.

4.2 The System Definition for Chemical Equilibria

Free energy, the tendency toward changes of state or the potential to do useful work, can be evaluated only for a well-defined chemical or mechanical system. In thermodynamic terms, a system is completely defined through the knowledge of the temperature, pressure, volume, and mass of all components present within the system boundary. These are the intensive and extensive thermodynamic variables which define the state of the system through application of the equation of state. The equation of state relates measurable physical quantities to the thermodynamic functions that define equilibrium.

Intensive Variables

Intensive variables describe the external, generally controllable, stresses on the system. Usually temperature (T) and pressure (P) are considered, but any energy quantity or field that affects the equilibrium state can be included in the equation of state and should be considered. Examples are magnetic fields and radiation fluxes. Temperature is the most straightforward measurement of the thermal energy available to the system and is expressed only using the "absolute" (or Kelvin) scale.

In a general equation such as the Gibbs relation above, temperature is the observable property linked to the thermodynamic "operator" entropy (S). Each atom in a chemical system is generally associated with about k_bT vibrational energy[2] which is readily available for atomic reorganization in response to other external fields or driving forces acting on the system. At room temperature, this corresponds to about 0.025 eV/atom or about 200 Joules/mol of atoms in the formula unit.[3] Atomic migration energies are on the order of 2000 to 10000 Joules/mol of atoms.

Pressure, P, is commonly (and wrongly) taken to represent either or both external mechanical pressure or vapor pressure of one or more components; these factors are both related to the system free energy through the total differential of the free energy but are used much differently in the system description. The pressure variable in the system description is the externally applied (mechanical) pressure.[4]

Positive mechanical pressure results in compression, causing atoms to move and thus affecting the bonding enthalpy and, through enthalpy, the heat capacity

---

[2] k_b is the Boltzmann constant = (gas constant, R)/(Avogadro's number, N_A).

[3] eV = electron volt or 96,490 coulombs/N_A (the charge on one electron) in a potential of 1 V. An eV is a convenient unit when talking of atomic scale processes. It is equal to 1.6×10^{-19} Joules.

[4] The external or mechanical pressure relates to the first "PV" work term whereas the partial pressures of the vapor species are related to the values of the chemical potentials, or partial molar free energies of the components, in the system.

(C_p) of the system. The change in volume also restricts or shifts the available electronic states (density of states) and so changes the internal energy and C_v. The heat capacities are defined for either constant volume or constant pressure conditions and are related via the thermal expansion coefficient and the bulk compressibility, $C_p - C_V = \frac{VT\alpha^2}{\beta}$, where

$$\alpha = \frac{1}{V}\left(\frac{\partial V}{\partial T}\right)_P$$

and

$$\beta = \frac{-1}{V}\left(\frac{\partial V}{\partial P}\right)_T.$$

The Chemical Components

The maximum number of unique components will be the number of chemical elements. The stated components must be appropriate for the description of each and every phase of the system. A chemical system's phases are uniform regions within the system with definable and homogeneous properties. Condensed phases, solids and liquids, are relatively close-packed assemblies of atoms and molecules. Whereas mechanical components can be arranged at will, the chemical elements cannot be arbitrarily mixed in the condensed oxide phases.

Earlier discussion has dealt with the unique nature of ionic solids, and the relatively stable, low atomic mobility, solids that result from the need to satisfy both electrostatic attractive atomic interactions and repulsive forces. Stoichiometry is a constraint on the definition of the components in oxide phase equilibria. Even in the vapor, certain groupings of atoms, molecules, or coordination groups are particularly persistent in ionic and oxide systems. If the system includes a vapor phase, the partial pressures are direct measures of chemical activity or partial molar free energy of the components.

The component descriptions—elemental, molecular or stoichiometric groups—must be appropriate to describe the composition of all phases and all equilibrium reactions among the phases of the system.

The System Boundary

A thermodynamic system must have a definable and appropriate boundary. Such a boundary may or may not correspond to any physical feature of the system or its containment, but must serve to effectively limit the portion of the universe under consideration. Classical thermodynamics requires that any transfers of mass or energy between the system and its environment be limited by a virtual partition, well guarded by Maxwell's demon, if the process is to be defined for reversible equilibrium. The virtual partition has no true physical counterpart, but can be defined as isolating the system in many cases.

The practical description of a laboratory system as closed or open must be done carefully. A common (and often incorrect) assumption in the description of condensed oxide phase equilibria is that the system is closed and that only the number of oxides (number of elements in the condensed phases minus one) need be considered.

Ceramic products are most often prepared as self-supporting shapes for solid-state reactive sintering. In the laboratory, mixed powders are reacted in open or closed crucibles to study reaction. Alternatively, pressed pellets or prepared shapes may be reacted or sintered while packed in a powder bed. The partial pressures of the components and the degree of free exchange with the ambient atmosphere determine whether the system may be considered closed, in which case the phase diagram will apply, or open, with nonconservation of component mass. Where reaction equilibrium is determined through the activity of a vapor species such as oxygen or nitrogen, reaction equilibrium can be controlled externally through control of that species partial pressure, as in the packed powder bed. Equilibrium will be determined in this case through the equilibrium constant for reaction, and not the phase diagram, as mass is not conserved.

For oxides with limited volatility and minimal oxygen non-stoichiometry, the contents of an open crucible or self-supporting shape at $T << T_m$ may be considered as a closed system under the condition that P_{Vapor} is very small ($<<10^5$ Pa) for all components.

If any species has a measurable partial pressure or if the observable state of the system (i.e., the phases present or phase composition) has a detectable dependence on the partial pressure of a gas phase species, the laboratory system is "open." In such a system, mass cannot be conserved unless efforts have been made to first condense all species to a known mass within a sealed "bomb" or containment which is then the reaction vessel. When heated, the isothermal reaction conditions at constant volume are not well defined, as pressure varies with the volume of the vessel and the number of moles of gaseous reactants present until steady state or equilibrium is reached. A high pressure and constant volume reaction condition is actually not uncommon in ceramic systems during firing if, for instance, a metastable reaction product or liquid entraps a species that becomes volatile at a higher temperature.

The system and its phases must be unaffected by the nature of the containment for the system boundary to exclude the containment itself. A condensed chemical system "in air" can be considered a "closed" system if none of the solid-state species have significant or measurable[5] partial pressures. This state will usually

[5]In a time of increasing sophistication of measurement, a significant partial pressure would be on the order of 10^1 to 10^2 Pa or 10^{-4} to 10^{-3} atm. Depending on the system and species, this pressure might not be easily measurable or could be far above the detection limit. Far lower partial pressures may be associated with component activities and defect equilibria.

be obtained at a temperature not more than about 90% of the lowest stable or metastable melting point temperature in the system.

If volatility is small or nonstoichiometric responses to the ambient conditions are limited to the surface only, owing to the very low transport rates usually obtained in oxides, the "system" away from the ambient interface may be considered as "closed." In that case it is common practice to remove the surface and near surface by grinding before evaluation.

The system boundary placement in all cases, if the system or a portion of the system may be considered closed, must be unambiguous if the phase diagram is to be used.

Extensive Variables

Extensive variables express the "amount" of material present within the system boundary and are related to the total energy content of the system through the bond energies and internal energies characteristic of the phases. Classical extensive variables are volume and mass. In chemical systems, the unit of expression is the "mol," or 6.022×10^{23} (N_A) chemical units (usually molecular or formula units), and we speak of molar volume or molecular weight, etc. As each atomic bond has an energy on the order of a few eV (electron volts), the bonds may be readily summed to yield an approximate value for the internal energy of the system at 0 Kelvin.

It should be noted that the ratio of two extensive variables results in an intensive variable. Examples are density (mass/volume) or tabulated molar quantities.

The volume of the system is fixed by the physical or arbitrary system boundary. A common assumption is that the dimensions of the system are the same for its thermodynamic consideration as are measured from its physical extent. In reality, a real chemical or material system is heterogeneous on several scales of observation, both macroscopic and microscopic.

A macroscale "system" may be considered as composed of many interacting local subsystems or "cells" and characterized by many local equilibria. The extent, composition, and thermodynamic "boundary" of each cell are generally not fixed over all ranges of temperature and pressure and the boundaries separating the local cells are open, that is, allowing free transfer of matter and energy. The approach to the system equilibrium state will be dictated by the kinetic processes linking transfer of matter and energy between the environment and individual cells or discrete phases and between the local regions of the material. Only after infinite time very near the equilibrium condition will the subsystems or "cells" become statistically homogeneous. The cell is an important concept in the Gibbsian view of the chemical system, where equilibrium is defined as a state in which the behavior of an individual cell is exactly predicted by the known state variables.

As chemical equilibrium is achieved through the motion of atoms and the time of observation is seldom infinite, a reasonable consideration is to establish subsystem dimensions using the known or estimated transport properties of the atomic species in their *slowest* transport paths. These characteristic chemical transport properties of the system, with a reasonable time variable, are used to define a transport length, similarly to the method commonly used to determine whether an "infinite slab" solution of the diffusion equation is appropriate. In diffusion studies, an infinite slab geometry allows simple interpretation of concentration data, as one end of the "system" is unaffected by the concentration gradient established through equilibration of the other end. It is found that a dimension $x \geq 4(Dt)^{1/2}$ satisfies this condition. For equilibration, this means that a characteristic dimension somewhat less than $(Dt)^{1/2}$ defines the extent of the equilibrating system truly under observation. For a experimental time of 24 hours (86,400 s) and diffusion coefficient of about 10^{-10} cm^2/s (10^{-14} m^2/s), the characteristic dimension is then about 30 μm in the solid phases.

Transport rates through the liquid are often on the order of 10^{-6} cm^2/s (10^{-10} m^2/s) or higher, so that the characteristic distance becomes about 3 mm. In low-viscosity systems, homogenization via convection is probable and a characteristic convection cell volume will determine the minimum meaningful system dimension.[6]

Significant transport through the vapor phase correlates with high vapor pressures of one or more components. Although this transport path is very inefficient for equilibration with or among condensed phases, a ceramic system that includes a significant vapor phase must be contained by a physical boundary that defines the system volume.

4.3 State Function Measurement or Calculation

For a system comprised of simple oxides or the elements for which the state variables are simply stated, the energetics of the system are theoretically fully determined. The state functions are then known or are calculated from the defined reference states. In these simpler chemical systems, tabulated data or possibly the phase diagram will be used to determine the physical nature of the lowest energy state. For complex systems, experimental measurement or calculation[7] of the equilibrium state and its relative energetics will more often be necessary.

---

[6]Convective transport can occur only under conditions of major inhomogeneity of the system: large density or thermal gradients. Therefore, although convective transport may be important in real molten or gaseous systems, such systems are far from equilibrium conditions which would allow equilibrium thermodynamic analysis. Many real systems are initially homogenized through a brief period of free convection followed by a longer period of quiescent homogenization.

[7]Calculation methods are truly outside the scope of this book, but are rapidly becoming a viable technique for the practical investigation of complex systems due to rapid advances in computational speed and power. A caution in application of calculated energy functions is that most calculation is performed under conditions of constant volume, whereas most experimental work is performed under constant pressure conditions. The internal energy (E) is most directly related to the Helmholtz free energy.

The most straightforward means to determine the difference in energy between states is to apply the "law of mass action" and measure the equilibrium constant of reaction to form the alternate states from components in their standard states as a function of temperature. The equilibrium constant is related directly to the free energy of reaction via the equation, $\Delta G = -RT \ln K_{eq}$. The enthalpy of reaction is determined by the measured equilibrium constant at various temperatures via the Van't Hoff isochore, which makes use of the Gibbs–Helmholtz relation between free energy and enthalpy. The relationship is stated

$$\frac{d \ln K}{d(1/T)} = \frac{-\Delta H_{rxn}}{R},$$

where R is the gas constant.

Effective application of the law of mass action for determination of free energies of formation requires that full equilibration can be achieved without kinetic restriction. Application of the Van't Hoff isochore requires that equilibration be possible over a range of temperatures. This restriction generally limits application of direct measurements of the equilibrium constant to vapor phase, solid–liquid, and some solid–vapor reactions.

Experimental reaction onset temperature and approximate reaction enthalpies can also be determined using calorimetry in many cases, if the reaction can be suitably contained. Calorimetry can also be used to measure heat capacity, and so to determine enthalpy and entropy changes with temperature in a system maintained at equilibrium. Dilatometric measurements of thermal expansion and compressibility are also related to the enthalpy and internal energy, as discussed earlier.

A reaction having a positive ΔH_{rxn} is called endothermic, heat energy will be consumed in the process of reaction. Similarly, a process having negative ΔH_{rxn} is called exothermic, meaning that heat will be evolved during reaction. An exothermic reaction's rate will often be limited by the transport of heat away from the reaction zone. Small sample sizes and carefully considered geometries are required for experimental measurement of reaction enthalpy.

At equilibrium $\Delta H_{rxn} = T\Delta S_{rxn}$. Classical thermodynamic treatments do not distinguish between heat energy that is involved in the disruption and reformation of the atomic lattices for solid-state reaction to proceed, the energy that is required for atomic transport or diffusion, and the thermal energy that is "lost" in the vibrational and electronic states (and may therefore be radiated to the environment). The energy to move atoms through the sublattice structure characteristic of oxides is substantial and constitutes the principle "activation energy" responsible for the kinetic limitation on the attainment of equilibrium. Energy lost to vibrational modes contributes to increased entropy and defect formation, necessary for atomic transport. Related electronic defect formation promotes bond reformation. The partitioning of heat energy as vibrational, electronic, and configurational entropy is not measured directly by calorimetry, but is important in understanding the progress of a reaction and the limits to attainment of phase equilibria in real systems.

Vibrational entropy is the probable distribution of energy over available vibrational modes, generally following a Boltzmann distribution. The configurational entropy is related to the number of ways atoms may be distributed over available sites or configurations, described through the "canonical ensemble" partition function, "W." The configurational entropy is then stated in terms of the partition function as $S_{config} = k_b \ln W$.

A partition function calculation is most simply demonstrated for random solutes or defects, where N sites are available for occupancy by n particles, distributed randomly.

$$W = \frac{N!}{(N-n)!n!} = \text{number of equivalent configurations}$$

For a cubic centimeter of a typical solid phase, N is taken as the number of atomic sites, or about 10^{22}. The number of solute particles for at 10 ppm impurity is then 10^{17}. These are very large numbers, so Sterling's approximation for the factorial terms is applied.

For large n, Sterling's approximation is $\ln n! = n \ln n - n$. As $N >> n$, $(N-n) \approx N$, so the configurational entropy is approximated

$$S = k_b \ln\left(\frac{N!}{(N-n)!n!}\right) \approx n \ln (N-n) - n \ln n$$

The calculated entropy is relative to the ideal lattice as reference state.

Calculation of the partition function for periodic and nonperiodic structures can be performed. Although this simple example demonstrates the idea of the calculation, it is not trivial to calculate or measure the entropy change of real structures to include defects. For the chosen reference state of the perfect crystal, an excess entropy can be readily determined using the simple approximation demonstrated. In an amorphous material or a liquid, the reference state is that of perfect randomness with no correlation of structure above the nearest-neighbor level in most cases. Defects are then regions having higher levels of atomic correlation (clusters) or regions of significant variation from average density. The partial molar entropy change for defect incorporation may be obtained by differentiation of the calculated partition function with respect to "n."

4.4 Solution Properties

Statistical treatments are well applied to the thermodynamic description of solutions in the solid and liquid states. The state functions of solutions and their behavior can also be described using more traditional terms. The correspondence between traditional and statistical thermodynamic properties of solutions gives a solid conceptual link between traditional and atomistic treatments of condensed systems.

The concept of the ideal solution is best applied to liquids and gases, systems without periodic structure that can be considered to be truly random mixtures. The ideal solution is composed of noninteracting particles. As they are noninteracting, the process of mixing volumes of species A and B results in a simple additive total volume ($v_A + v_B = V$) and there is no change in either internal energy or enthalpy on mixing. The behavior of inert gases at very low pressures is reasonably "ideal," as the mean free path is long and the probability of random interaction via collision is small.

Ideal solution behavior is often extended for the description of behavior in crystalline materials, particularly for the behavior of minor impurities and defects which can be treated statistically as random. At ppb levels, defects and solutes are treated as noninteracting species that just happen to exist in a homogeneous dielectric medium. This is reasonably consistent with the traditional treatment of ideal solutions, where the solute has a vanishingly small solubility in the solid solvent and the behaviors of the solute or solvent are independent.

"Ideal solution behavior" in condensed phases is usually found in the "dilute solution limit." As relative density increases, the concept of noninteracting particles becomes limited by the nature of the solute and the physical and dielectric properties of the solvent. As a rough guide, the probability of solute atoms or particles occupying near-neighbor positions when randomly distributed in the solvent must be less than about 1% for "zero" interaction. In a 10 mol% solution, the probability of near-neighbor occupancy in a condensed phase is almost 100%. The dilute solution limit in condensed phases is about 0.1 mol% solute, although it is found that the properties of solutions having up to 10% may be simply described as ideal for many purposes[8] when solute and solvent properties are similar and there is no expectation of reaction. Additivity in molar volume is a good indicator as to whether the solution may be approximated as ideal.

The properties of ideal solutions are described in terms of the behaviors of the solvent (A) and the solute (B). The solvent (dominant component) is generally seen to obey Raoult's law behavior: $P_A = x_A P_{A'}$, where $P_{A'}$ would be the partial pressure of A usually observed over the pure phase at the experimental temperature and pressure. As the solvent is considered ideal, $P_{A'}$ can be calculated if $\Delta G(T,P)$ is known for the pure solvent. Conversely, the impurity concentration, $1 - x_A$, may be calculated from the measured boiling point temperature.

The properties of the dilute solute (minor component) in solid or liquid solution

[8]For a substitutional cation solute in an oxide, adjacent cation lattice sites are separated or "shielded" by the intervening oxygen ion. The shielding effect limits interaction and is probably responsible for the higher effective limit for observation of "dilute solution behavior" in these solid solutions.

are most often described using Henry's law: $P_B = x_B K_B$, where K_B in this context is a constant experimentally determined to be valid over a range of values of x_B, generally as $x_B \rightarrow 0$. Henry's law constants are specific to the solvent system for each solute and cannot be calculated directly from the thermodynamic properties of either phase unless a degree of interaction is proposed.

The partial pressure of a component is directly related to the partial molar free energy or chemical potential of that component in the solution. In the above description, the linear behavior ascribed to partial pressure of the solvent with composition could as well be ascribed to partial molar volume or other first-order property and then directly related to the variance of the chemical potential. In solid solutions, the linear change in first-order properties such as lattice parameter or molar volume is referred to as "Vegard's law behavior" or simply "additivity."

Liquid Solutions

The colligative properties of liquids are direct consequences of the assumption that the solution is Raoultian. It is assumed that the solute is not volatile (its concentration therefore does not change with time) and also that the solute is not soluble in the solid solvent. The latter assumption is most important as it provides the basis for the use of the pure solid and pure liquid solvent as the thermodynamic reference states for the solution thermodynamic properties.

A common application of the colligative property relationships that is relevant to phase equilibria is through application of the "freezing point depression" relation to estimate the liquidus curve. If melt enthalpy is known for the pure materials, then

$$\ln x = \left[\frac{-\Delta H_{\mathrm{melt}}}{RT}\right]\left(\frac{1}{T} - \frac{1}{T_{\mathrm{m}}}\right)$$

where B is the solute and all other properties refer to the solvent. As the thermodynamic reference state for this relationship is the pure solid, it can only be used to estimate the liquidus temperature near the melting point of the pure material. Obviously, the solubility curve or liquidus can be used similarly to estimate the enthalpy of melting of the pure phase, as the tangent to the actual liquidus should approach the ideal slope as $x_B \rightarrow 0$. A similar relationship can be written to express the change in transition temperature for any condensed phase transition near the stoichiometric composition on the introduction of an impurity.

Statistical Treatment of Solutions

A result similar to the freezing point depression can be obtained through application of statistical methods in the calculation of the properties of the solution. Statistical methods are particularly useful for multicomponent solutions and glasses. As noted above, solute and defect distribution in the crystalline state can be treated statistically, allowing calculation of the statistical entropy of both liquid solution and solid solution. In an ideal solution, the heat of mixing is zero, so that the variation of the statistically derived entropy change on mixing relates directly to the change in free energy and the variation in the chemical potential. Phase compatibility is then determined through direct solution for the condition of equivalent chemical potential of the components in all phases, possible only if a common reference state is chosen for calculation of all phases' properties.

The ideal solution model used in entropy change calculation assumes that there is no interaction between atoms or molecular species, or at least none that is outside the average bond enthalpy seen for the individual components, $(H_{A\text{-}A} + H_{B\text{-}B})/2 \approx H_{A\text{-}B}$ and that the entropy of the system may be described using statistical means through a simple partition function. For a simple two-component system, the statistical entropy of random mixing is simply stated

$$\Delta S_{\text{mixing, ideal}} = -R\ [(x_A \ln(x_A)) + (x_B \ln(x_B))]$$

The entropy calculated is per mol of particles, $x_A = 1 - x_B$ are mole fractions.

To determine graphically the liquid composition in equilibrium with the "pure solid" at $T < T_m$, the free energies of both liquid and solid are expressed with reference to a common state, usually the pure liquids as a mechanical mixture. At T not equal to T_m, the relative free energy difference is calculated using the relationship $G_{\text{solid}} - G_{\text{liquid}} = \Delta H_{\text{fusion}} \ln(T_m /T)$, derived using the condition at equilibrium that

$$\frac{\partial(\Delta G)}{\partial T} = -\Delta S.$$

This value describes a point on the common tangent to the free energy–composition functions of both solid and liquid, defining the composition at the condition $\mu(\text{solid}) = \mu(\text{liquid})$.

Either of the traditional thermodynamic "ideal" or the statistically "ideal" cases implies that the solution will be perfectly random, with no tendency toward ordering or compound formation (in the case of solid solutions), and that the enthalpy or heat of solution will be zero. As the assumptions are identical, it is not surprising that the results are usually the same via either statistical or traditional means for ideal solutions.

A deviation from ideal values, actual *minus* ideal, is termed an "excess" quantity. Many solution models are based on the "regular solution" model in which the excess enthalpy is nonzero, recognizing the different energies of interaction between real particles, but the excess entropy is zero, meaning that no ordering exists. Regular solution behavior can be readily incorporated into the statistical treatment of solution behavior to allow meaningful treatment of real systems and the phenomena of solution instabilities such as phase separation. To account fully for the behavior of real systems it is usual to include terms that account for local interactions in the statistically described system, and solve for the energetically compatible phase compositions directly using the simultaneous equations.

Solid Solutions

The simplified treatment of complex solid-state systems is through their relation to a perfect or ideal crystal reference state and calculating the "excess" quantities associated with random defects or alternative ordered arrangements of defects. Real behavior, or a deviation from statistical randomness, can be described as an "excess" relative to the statistically random ideal.

Thermodynamic behaviors and "excess quantities" due to solution formation can be calculated statistically, while the thermodynamic properties of the ideal crystal, glass, or liquid state may remain unknown. Thus, defect and solution behavior has been extensively treated in the literature, and with good success, while the theoretical description of pure condensed states is less advanced.

For all systems, the free energy change ΔG is a measure of the tendency for change between initial and final states. Path independence of the state functions allows real processes to be idealized and analyzed, whether thermomechanical or electrochemical processes are considered. For many systems, a state that is far removed from equilibrium is so unstable an "approach to equilibrium" cannot be always observed because of its extreme rapidity. The idealized reversible process required to describe such a process need not relate to any physical process of mass or energy transfer.

This rapid and certain approach to equilibrium is contrasted with the behavior of real oxide systems, where the approach to equilibrium is predicated on local reaction and atomic transport to continue reaction once initiated. A stable multicomponent oxide system is characterized by the continuity of the chemical potential functions of all components throughout the phases of the system, which is achieved through reaction, and minimization of the spatial gradients in the chemical potential functions. The approach to equilibrium from its stable, but nonequilibrium, state cannot be separated from the physical processes of heat, mass, and electronic transport that characterize its phases.

Traditional thermodynamic treatment also does not include the contributions of surfaces or interfaces to the total energy and the chemical potential gradients. As previously discussed, these contributions can be large in ionic and oxide systems. Nanoscale observations, particularly in polycrystalline samples, cannot necessarily be taken as representative of the macroscopic system unless all local field effects are carefully accounted for.

The bridge between nanoscale observations and thermodynamic description of the ideal, homogeneous chemical system is the statistical description of quasi-chemical defect structures and their energetics. More exacting energetic descriptions of electronic and magnetic contributions to system stability and equilibrium may require quantum mechanical treatments, well outside the scope of this book but well treated in specialist sources.

4.5 Congruent Phase Transition

A change of state—melting, vaporization, or condensation of a liquid or crystalline solid—is a fundamental first-order phase transition of an equilibrium homogeneous system. Phase transformation is a *congruent* process, occurring without compositional change. The free energy function is continuous throughout the transition; the variance in its derivatives define the "order" of the transition. The driving force for phase transformation is the change in free energy per unit volume, dG/dV, of material transformed.

The large observable volume change corresponds to an abrupt change in the state functions, particularly internal energy and entropy, which are fundamentally related to the phases' characteristic atomic arrangements or symmetries. Two phases, having the same compositions but different structures, can coexist at equilibrium under the condition $\Delta G_{(1\text{-}2)} = 0$. In an homogeneous system, congruent transition is a univariant process, meaning that the equilibrium values of temperature and pressure are not independent of each other, but rather are related through the Clapeyron equation.

The Clapeyron equation is derived by setting equal the differential Gibbs equation for each phase of unchanging composition, $\partial G_1 = V_1\partial P - S_1\partial T$ and $\partial G_2 = V_2\partial P - S_2\partial T$, as both the free energy and the free energy gradients of the two equilibrium phases are equal at the transition.

Thus, for melting at T_{m}, P_{m}

$$\frac{dP}{dT} = \frac{\Delta S_m}{\Delta V_m};$$

$$\text{Solid} \rightarrow \text{Liquid} \qquad p = p' + \frac{\Delta H_m}{\Delta V_m} \ln \frac{T}{T'}$$

$$\text{Liquid} \rightarrow \text{Vapor} \qquad \frac{d\ln p}{dT} = \frac{\Delta H_{\text{vaporization}}}{RT^2}$$

These relationships are of significant practical importance when considering the effects of process or environmental conditions on equilibrium phase transition. The Clapeyron equation and its associated forms for melting or vaporization are examples of a group of equilibrium relations that may be derived from various statements of the Gibbs equation. A practical set of these equations are referred to as Maxwell's relations, the various partial differentials that relate the thermodynamic functions in the Gibbs equation. All can be useful in deriving equations of state or to extract thermodynamic data from experimental observations of congruent phase transitions.

As a set, the Maxwell's relations are often stated;

$$\left(\frac{\partial T}{\partial V}\right)_S = \left(\frac{-\partial P}{\partial S}\right)_V \qquad \left(\frac{\partial T}{\partial P}\right)_S = \left(\frac{\partial V}{\partial S}\right)_P$$

$$\left(\frac{\partial P}{\partial T}\right)_V = \left(\frac{\partial S}{\partial V}\right)_T \qquad \left(\frac{\partial V}{\partial T}\right)_P = \left(\frac{\partial S}{\partial P}\right)_T$$

Polymorphic Phase Transformations

It is customary to speak of *polymorphic* crystal forms of the solid phase and *allotropes* when speaking of different states of matter having the same composition. An allotropic transition will always be a first-order transition. Polymorphic transformations may be first-order, second-order, or "lambda-type" transitions.

In a first-order transition, the state functions (volume, entropy, enthalpy) each exhibit abrupt step-function change at the critical point. These functions are the first derivatives of the free energy function with respect to the intensive variables. The derivative of the chemical potential is continuous but shows an inflection and change in slope at the critical point. The second derivatives of the free energy, most notably the heat capacity, approach infinity at the critical point.

Second-order transitions are then theoretically those that have continuity in first-order state functions throughout but exhibit step function change in second-order properties. As measurement techniques become increasingly sophisticated it has been shown that most supposed second-order transitions do not show a true step-function variation but rather are lambda-type transitions.

In a "lambda" transition, the first-order properties are continuous and the second-order properties show step-function behavior in routine measurement. An example is sketched in Figure 4.2. On close examination, all functions and their derivatives exhibit point-to-point continuity but an extremely small step behavior is seen in first-order properties. The name "lambda-type" refers to the shape of the second-order property curve when carefully measured, which is seen to go through a sharp, high maximum or singularity at the transition temperature, such that the curve looks rather like a script "lambda." Examples of lambda transitions are found in ferromagnetism, superfluid–normal fluid transitions, and order–disorder transformations in alloys.

Magnetic transitions are detected readily in the heat capacity, principally owing to the electronic contribution to the lattice entropy, as well as in the magnetic properties. These are order–disorder transformations and have varied characteristics owing to the existence of multiple sublattice and the collective exchange interactions possible. In a ferromagnetic material, the spin alignment is gradually

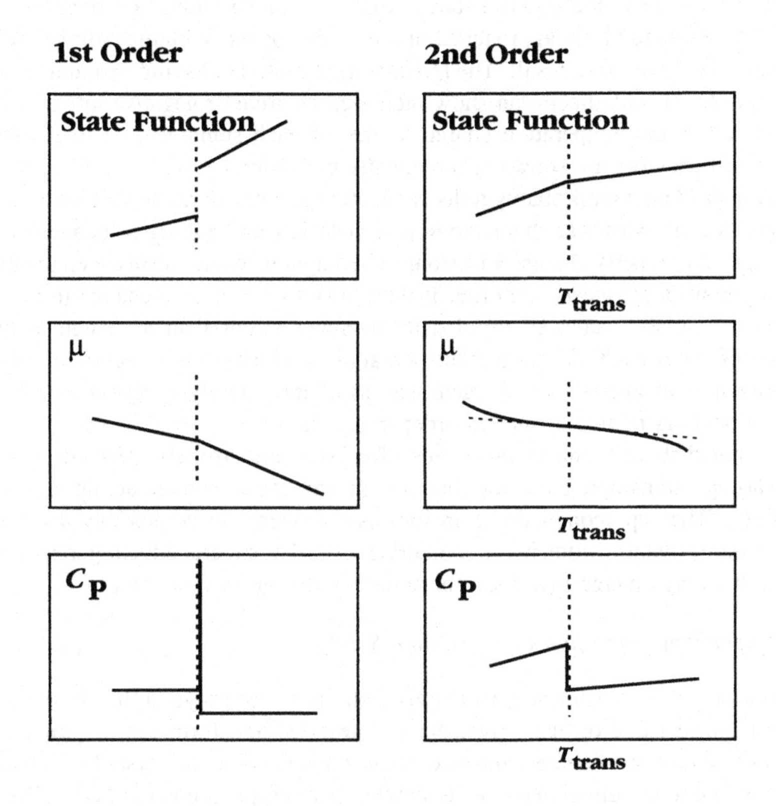

Lambda-type behavior

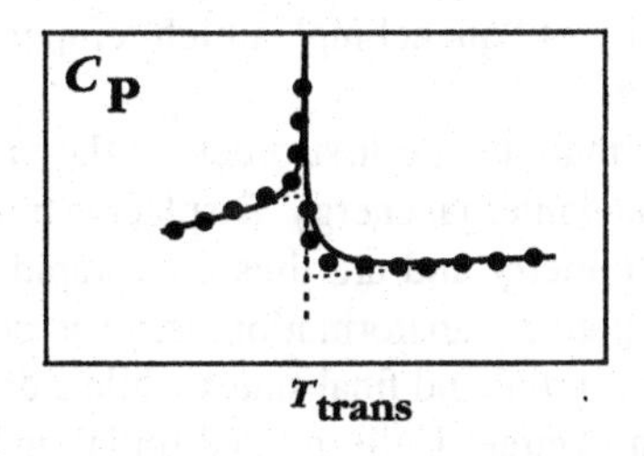

Figure 4.2 Schematic examples of first-order, second-order and "lambda"-type transition behaviors.

randomized as T_C is approached and a gradual drop off in saturation magnetization is observed. The saturation magnetization of an antiferromagnetic material, having zero net moment in the ordered state, exhibits "lambda" transition behavior as the Curie temperature is approached and disorder of the sublattice allows high exchange interactions to result. The ferrimagnetic material, having partial interaction of ordered sublattices, can show increased positive or negative interactions depending on the temperature characteristics of each sublattice. Characteristic transition behaviors are shown schematically in Figure 4.3.

"Lambda" behavior probably reflects the atomistic processes and local compositional uncertainties that characterize real materials undergoing a second-order type physical property change. Electronic and magnetic phenomena are characteristically considered as second order if the transition involves a change in occupancy of available states in the density of states rather than a change in the density of states itself.[9] There remains a real uncertainty in the detection of these phenomena and also in the characterization of the underlying symmetrical or tensor operators related to second-order transition.

It is noted that the tensor properties of crystals are strongly affected during lambda-type transition behavior, but not during the superconducting–normal transition. The superconducting transition is considered to be possibly the only true second-order transition in the crystalline solid. The superconducting transition is also the only change that does not require a change in symmetry.

Congruent Transformation in Crystalline Solids

First-order phase transformations between allotropes or polymorphs are termed either reconstructive or displacive. A reconstructive transformation requires the coordinated motion of nearest and next-nearest neighbors in the crystalline lattice, therefore the transformation proceeds slowly. In allotropic transitions, crystallization of a solid from the liquid phase is a first-order reconstructive transformation. The rate of the transformation is kinetically limited by the phenomena of nucleation and growth in either solid–solid or liquid–solid transformation. The kinetic limit results in the possibility of "quenching" a high-temperature phase liquid or crystalline condensed phase.

All polymorphic transformations are first order, as they are associated with a change in unit cell volume and internal energy. Displacive transformations require only a minor change in symmetry and are thus quite rapid—often termed "unquenchable." In a reconstructive transformation, there is no necessary link between the symmetries of the initial and final states while a displacive transformation is necessarily between symmetrically related initial and final states.

Order–disorder transformations sometimes fall into a difficult area of classifi-

[9]The density of states is directly related to the system volume, so a change of state that results in a change in the density of states must be first order.

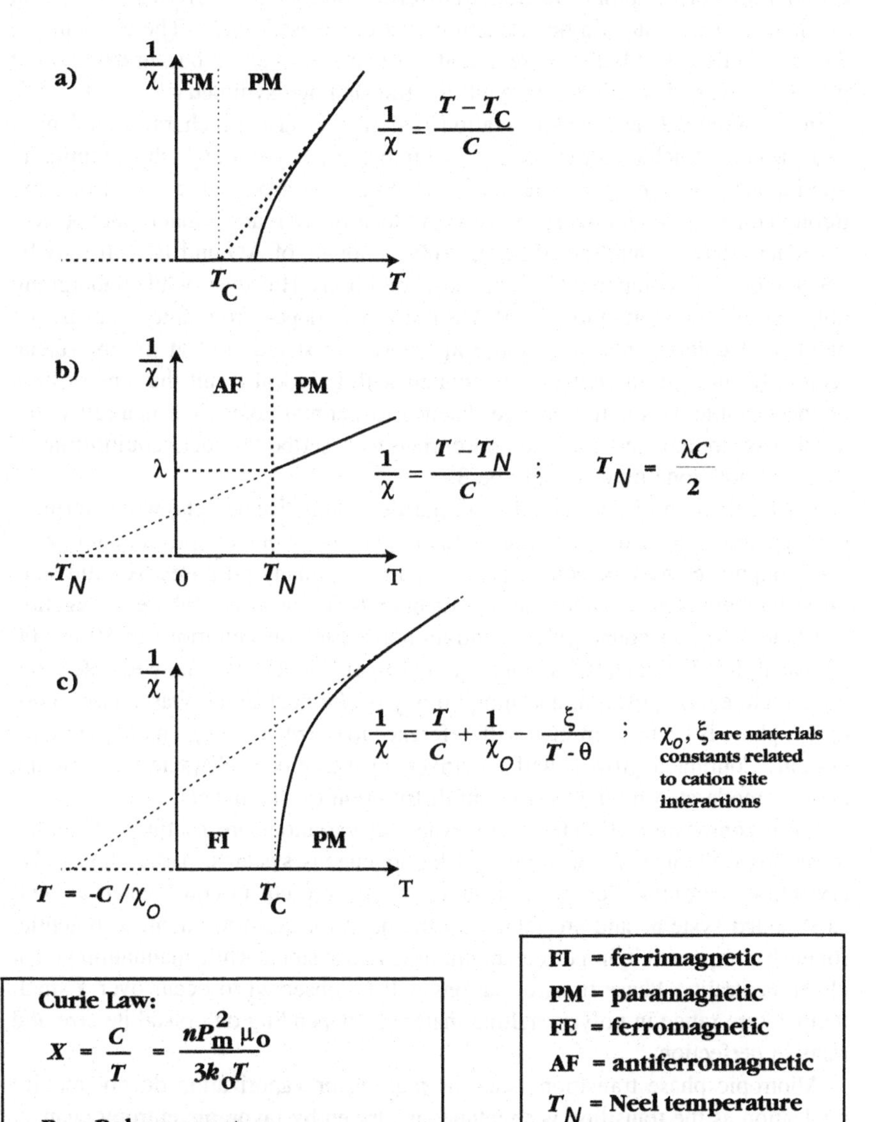

Figure 4.3 Schematic transition behavior in (a) ferromagnetic, (b) antiferromagnetic, and (c) ferrimagnetic materials.

cation. Polymorphic order–disorder transformation usually involves the formation or disruption of a superlattice structure in a condensed phase. The elements of the superlattice exist both before and after the transformation but, in the ordered phase, are related or aligned through the translational symmetry.[10]

In a compound, any transformation region may also be characterized by a relaxation of stoichiometric "rules." The transitional system, which is nominally n-component, where n is the number of chemical elements reduced by 1 or more through rules of stoichiometry, may behave locally in a manner that is unexpected. Consider ABO_3, composed of equal molar amounts of AO and BO_2. It may be considered "one-component" under most conditions. However, while undergoing polymorphic transformation, local A/B ratios may depart from unity, causing the number of components to be more appropriately stated as 2 at the interfacial region. If under strain, defects are created with the local result that one or both of the metallic ions could change valence to maintain electrical neutrality, the number of components that are appropriate to describe the local equilibrium of the transition zone may actually be 3.

The local composition (including impurities and lattice defects) will determine the local gradients and driving forces toward change. For example, a system with 0.1% impurities may be considered as "pure" for many purposes. A solid-state phase transition tends to nucleate at grain boundaries; however, where segregation coefficients (grain boundary concentration/average concentration) of 10 to 100 are usual. It is the boundary composition (1 to 10% impurity) that will determine the structure, composition, and morphology of the nucleating phase. The nature of this phase is best determined with reference to the phase diagram of the impure system. Continued growth will be driven by the global characteristics of the system, moderated by the nature and distribution of the nuclei.

Local compositional, defect, and structural variations are routinely observed in microscopic analysis of nominally homogeneous systems, particularly polycrystalline ceramics. Composition gradients are present in equilibrated as well as stranded systems and are related to the need for uniform chemical potential for each component in imperfect, noninfinite real systems while maintaining local electroneutrality. Phase transformations will be observed to occur over a small temperature range in polycrystalline materials, depending on crystallite size and relative perfection.

Allotropic phase transitions such as melting or vaporization do not require nucleation as the transition is spontaneous, driven by favorable entropy change. Phase transitions on cooling, however, are not spontaneous and must generally

[10]True order–disorder transformation should not be confused with "ordering" transformations in chemically complex phases. Commonly, these ordering phenomena are observed in phases that are not in an equilibrium state owing to rapid changes in temperature, pressure, or chemical environment ("quenching"). When the system is thus trapped or "stranded" in a nonequilibrium state, further transformation is not describable in terms of equilibrium thermodynamics of transformation.

be nucleated. Nucleation phenomena and the effects on solid-state transformation are most readily described using the statistics of simple probability.

4.6 Nucleation Phenomena and Transformation Kinetics

Observability, Persistence, and Equilibrium

A system can exist in many physical states, most either metastable or unstable. The equilibrium state has the lowest energy. An unstable state has a very low probability of being observed, a metastable state may have a reasonable probability of being observed under certain circumstances, whereas the equilibrium state can be achieved reversibly and is observed to have consistent properties. The free energy of possible physical states may be estimated by comparison with an apparent or known equilibrium state.

The probability of observation of a given state is related to its energy generally through the Boltzmann distribution:

$$P_j = e^{-\left(\frac{E_j - E_0}{k_b T}\right)}$$

A state "j" having energy significantly higher than the reference stable state, with energy E_0, has a very low probability of observation. A state with a more readily achievable atomic configuration relative to the initial state may be observed quite often, if it is also persistent. An example would be the case of metastable displacive transformation to a reduced, but not lowest, energy state, which could initiate and grow in preference to an equilibrium transformation that required long-range atomic reconstruction.

Equilibrium reconstructive transformations in solid oxides require reasonably high atomic mobility for equilibrium to be the favored observable state. The multiple sublattice structure, stoichiometry, and electroneutrality requirements of the ionic solid restrict free mobility of atomic species; some correlation will always be the case in condensed phases.

A system without sufficient energy to approach equilibrium has been termed "stranded" in the earlier discussion. A stranded system may possess large excess internal or free energy but requires significant perturbation to supply activation energy for transition. A metastable system will similarly require perturbation to initiate equilibrium transformation which then may progress freely, whereas a stranded system cannot generally progress toward equilibrium without continued inputs of energy.

The homologous or reduced temperature, $T_h = T/T_m$, is a significant guide to the energy requirement for unconstrained mobility for transition toward the equilibrium state. The homologous temperature is also probably the most reliable kinetic factor available directly from the equilibrium phase diagram.

The system's temperature is a measure of the average energy per particle. In a solid, this average is generally stated to be on the order of $3/2k_BT$, for the three vibrational degrees of freedom. The distribution of energies can be assumed as approximated by a normal distribution about the mean value, although the actual values are quantized. For a reaction to proceed, an atom or particle must have sufficient energy to change its position in the bonded structure. The probability of sufficient energy must be

$$\geq \exp \frac{(E(T)_{avg} - E(T_m))}{k_bT} \tag{4.1}$$

How large must this energy be? The material's melting point is a useful benchmark to establish a relative scale. At $T = T_m$, all particles are free to move as in the liquid, so intuitively it is clear that their average energy is sufficient to overcome any reasonable activation energy. The complete distribution of particle energies follows Fermi–Dirac statistics; however, the high energy "tail" may be approximated by a Boltzmann distribution of the form given in Eq. (4.1), where $E(T)_{avg} = 3/2k_bT$ and $E(T_m) = 3/2k_bT_m$.

Making these substitutions, Eq. (4.1) can be recast in terms of the material homologous temperature as

$$\geq \exp\left(\frac{3}{2}\left(1 - \frac{1}{T_h}\right)\right) \quad \text{where} \quad T_h = \frac{T}{T_m} \text{ (all } T \text{ in Kelvin)} \tag{4.2}$$

which is the probability of a particle having sufficient energy to move "freely."

A plot of Eq. (4.2) is given in Figure 4.4. It is readily seen that below $T_h \approx 0.3$, there is almost zero probability of any particle having this energy required for free movement, while at $T_h \geq 0.7$ there is a better than 50% probability of occupancy of this energy state. These are uncorrelated probabilities, so each of the n atoms required to form a cluster or prenucleus of the stable phase must have this energy. For a small cluster of $n = 3$, a 1% probability of formation requires that each atom or particle have a $0.01^{1/3}$ ($\approx$0.22) probability of sufficient energy. This condition occurs at appromixately $T_h > 0.5$. For $n = 10$, $T_h > 0.75$ is required.

$T = 0.7T_m$ is known in some disciplines as the Tammann temperature (see below), a temperature above which solid-state transformations can generally proceed. In ionic solids characterized by high defect formation energies, the Tammann temperature is closer to $0.8T_m$. In metallurgy, the temperature range 0.7 to $0.8T_m$ is known as the recrystallization temperature.

T–T–T Phenomena

Limits to transformation of either truly metastable or kinetically stranded systems are known as "time–temperature–transformation" phenomena, experimentally

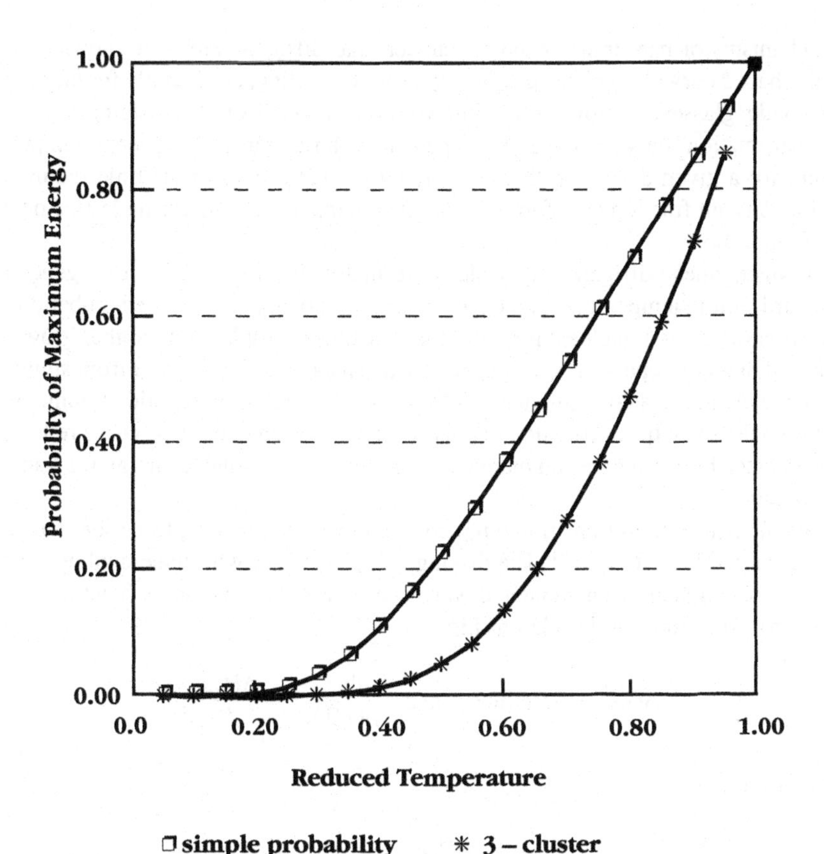

Figure 4.4 Plot of the probability of maximum energy using the normalized "homologous" temperature scale.

described by the T–T–T curve. The phenomenon of undercooling prior to the nucleation of the stable phase was noted by many early workers of phase equilibria and particularly by Tammann (1925).

In his experiments, a liquid was supercooled as droplets on an ultraclean plate; the occurrence of crystallization at known undercoolings was then noted. The apparent crystallization probability as a function of temperature described a bell-shaped curve, with a maximum occurring at a temperature significantly below the equilibrium value. Tammann attributed this behavior to the statistical nature of the nucleation event, saying "Spontaneous transformation is subject to the laws of probability, whereas non-spontaneous transformation in uniform substances is governed by the flow of heat and by a linear velocity dependent on temperature." The statistical behavior was taken as supporting the atomistic nature of matter, which was not firmly established at that time.

Today, the atomistic nature of matter is universally accepted and the many mechanisms of heat transfer, mass transfer, and diffusion are well known. The bell-shaped curve for crystallization versus undercooling, particularly for the case of oxide glasses, is now better known as a T–T–T or "time–temperature–transformation" curve, and is a plot of percent material transformed versus temperature for a given annealing time or may be plotted to shown the time required for a constant fraction transformed at a given temperature. An example is shown in Figure 4.5.

Longer times will generally yield more material transformed with a general upward shift in temperature of the curve maximum owing to increased probability of observation of a nucleation event at longer times, coupled with higher growth rates at higher temperatures. Its characteristic shape is easily derived from simple statistical and energetic arguments. T–T–T behavior is characteristic of homogeneous solid-state transformation in most systems. The characteristic curve reflects the balance between thermodynamic driving force and available energy for transformation.

At the equilibrium transition temperature, there is no driving force for change to go forward as $\Delta G_{trans} = 0$. The driving force increases with undercooling, and can be calculated from tabulated values of the thermodynamic functions (if known) using the simple relationship

$$\Delta G_{trans} = -RT \ln K \quad \text{and} \quad \frac{\partial \ln K}{\partial (1/T)} = -\frac{\Delta H_{trans}}{R}.$$

Integrating, then

$$\ln K = \frac{-\Delta H_{trans}}{RT} + (\text{constant})$$

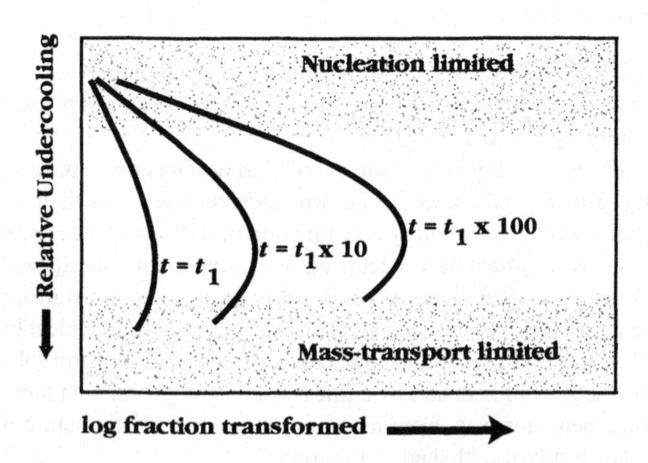

Figure 4.5 Characteristic T–T–T or "time–temperature–transformation" curve.

Substituting $\Delta G = 0$ at $T = T_{trans}$, the constant of integration can be solved and the result, the relative free energy change or driving force at $T < T_m$, can be expressed as

$$\text{Driving force} \geq \Delta H^0_{trans}\left(1 - \frac{T}{T_{trans}}\right)$$

The driving force, or free energy difference, increases with undercooling.

In Figure 4.6, the relative trends are diagrammed and combined. Immediately, the "bell-curve" is apparent. The combined effects of increasing driving force with a decreasing available average energy for random motion result in the characteristic curve shape. Increasing complexity of the nucleus, simulated here by demanding increased correlation, sharpens the curve but also depresses its maximum.

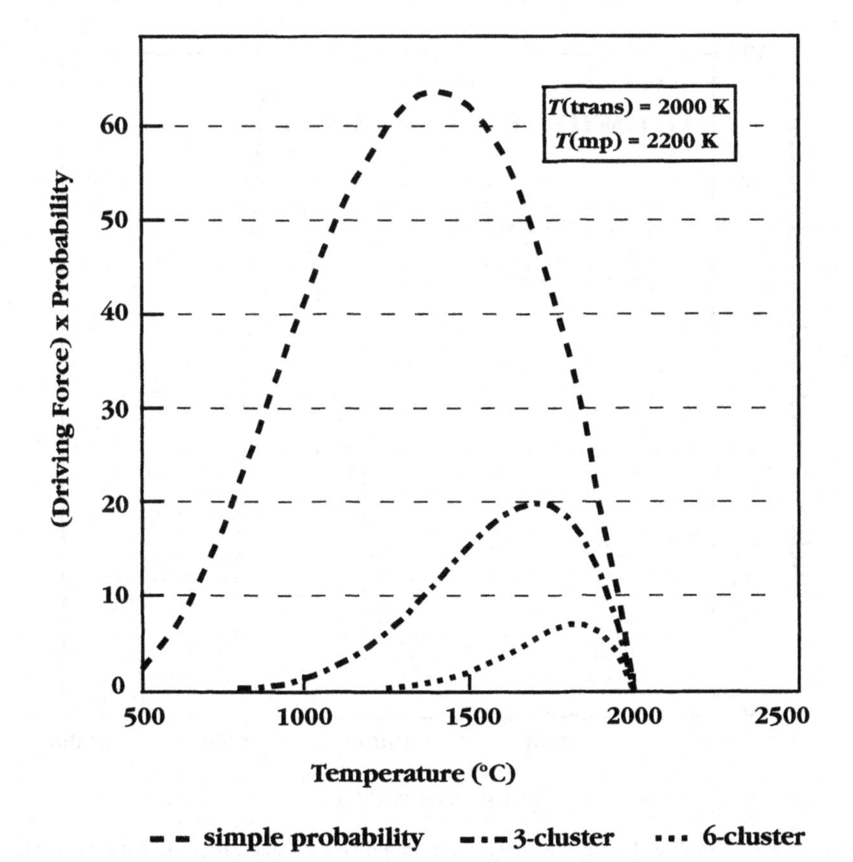

Figure 4.6 Schematic diagram of the relative probability of spontaneous fluctuation with thermodynamic driving force (undercooling).

Nucleation is governed by the available energy for random atomic motion and the statistical probability of formation of the "right" configuration. The driving force represents a relative probability that this configuration will persist. Only if the diffusivity of all species is sufficiently high, however, will the nucleus grow.

Atomic diffusion is also a thermally activated process with a characteristic dependence

$$D(T) = D_o \exp\left(-\frac{Q}{RT}\right)$$

The activation energy, Q, is characteristically large in comparison with RT, yielding a very steep exponential decrease in mobility with decreasing temperature. The combined effects are diagrammed in Figure 4.7.

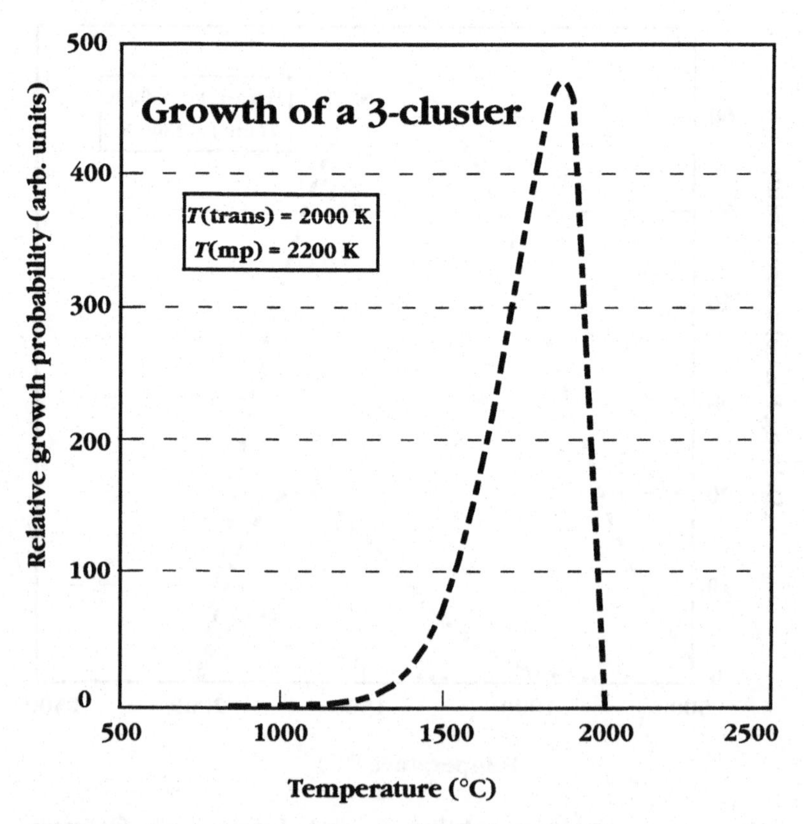

Figure 4.7 Combined effects of decreasing mobility with nucleation probability result in a sharply delimited window for transition in the undercooled system.

Heterogeneous Nucleation

The thermodynamic driving force described above is also the net energy available in the system to accomplish the work of transition. This work entails the physical motion of atoms and molecules, electronic transitions as bonding states are redistributed, and the work to create an interface between phases.

The simple statistics that yield the shape of the T–T–T curve cannot fully consider the constraints that are characteristic of real solid state transformations. Phase transformation in real materials requires a nucleation event that is more complex than simple statistical treatment will allow, even for simple metals where stoichiometry and structure are not significantly changed between parent and product phase. In an unconstrained system, nucleation may occur freely (homogenous nucleation) but is statistically unlikely for all the reasons outlined above.[11] More usually, a heterogenous "kick-start" is required.

Real solid-state systems are constrained and work required for a change of state includes not only the purely mechanical bulk strains, recalling that a first-order transition requires a net ΔV, but also the creation of new surface with its characteristic energy, γdA. In the early (or nucleation) stages of transition, the work required to extend the surface can easily exceed that available from the bulk transformation and growth cannot proceed without significant undercooling. Growth at or along an existing interface or in a highly defective region may be energetically neutral or even favorable through replacing a higher energy with a lower energy interface.

Heterogeneous nucleation is generally attributed to impurities or inclusions. Heterogenous nucleation can also be considered as nucleation that is occurring in a physical region of the solid where constraints are reduced owing to the existence of defects, impurities, or a preexisting interface.

For the nucleation of a new phase at an existing interface, a three-phase junction must be created. The probabilities of forming a stable low-energy nucleus of a new phase on a perfect crystal is low, simply because of the increased energy due to the formation of new surface and notwithstanding the statistical arguments against the nucleation event. It is not surprising that an amorphous or highly defective surface (or substrate) is seen to greatly enhance the nucleation rate of a new phase, as both $\Delta G_{formation}$ is increased and the nucleation barrier diminished.

The physical situation at the interface is traditionally diagrammed as in Figure 4.8. An energy balance between the surface tensions must be established that satisfies the listed criteria.

---

[11]Optimal radiative heat transfer and near-zero gravity conditions should favor spontaneous nucleation and growth of near-perfect crystals from the vapor phase, yet even snow crystals forming on a clear night are found to have nuclei of dust or soot.

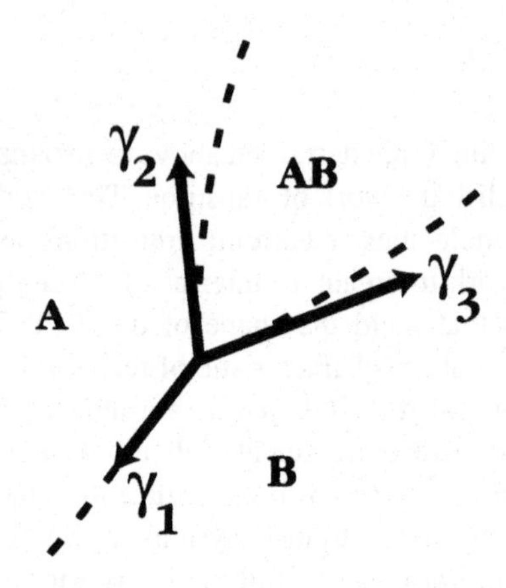

Figure 4.8 Schematic diagram of surface tensions at a three-phase junction.

$$(\gamma_2 + \gamma_3) > \gamma_1$$

$$(\gamma_2 + \gamma_1) > \gamma_3$$

$$(\gamma_3 + \gamma_1) > \gamma_2$$

where

γ_1 = A|B surface tension.

γ_2 = A|AB surface tension.

γ_3 = B|AB surface tension.

Satisfaction of these requirements on the inequalities is necessary to ensure shape stability of the nucleus.

The change in surface tension or surface energy on going from two to three phases can be written $\Delta\gamma = (\gamma + \gamma_3) - \gamma_1$. $\Delta\gamma$ is generally on the order of 1000 ergs/cm^2, and may be positive or negative. For $\Delta\gamma > 0$, critical nucleation behavior due to shape instability or critical thickness behavior may be noted. The latter is observed where the volume enthalpy change on reaction just balances the surface energy change:

$$\text{critical thickness, } d = \left(\frac{\Delta\gamma}{\Delta h}\right),$$

where Δh is the volume enthalpy change. (This is usually within a factor of two of the observed value.)

$\Delta\gamma < 0$ does not rule out nucleation but does indicate that nucleation is not stable at the original interface. If the constraints on surface tension are not satisfied, it may well be the case that nucleation will be observed away from the interface, on a free surface or pore. Note, however, that if nucleation is not stable at an interface, the nucleation rate will critically depend on the mass transport rates across the interface and impurity solubility through the "host" phase to the nucleation site.

The nucleation rate of a new phase at an existing two-phase interface is found to have a form that is governed by the balance between bulk (reaction) and surface free energies, as well as the activation energy for diffusional transport to and across the interface. The nucleation rate, N, can be related to the surface and diffusional activation energies

$$N \propto \exp\left(\frac{-\Delta G^*}{k_b T}\right) \exp\left(-\frac{Q}{k_b T}\right),$$

where

$$\Delta G^* = \frac{\Delta\sigma^3}{\Delta G_{\text{trans}}{}^2}$$

and $\Delta\sigma$ is the surface energy change.

Nucleation rate limited growth is most often observed where the volume enthalpy of transformation, Δh, is smaller than 400 J/cm^3. (In discussions of constrained nucleation, as at an internal interface between solid phases, the Helmholtz free energy may be more appropriately used due to the existence of a fixed reaction volume.)

Transformation rates are generally improved in defective oxides, owing to increased heat and mass transfer.[12] However, highly impure materials (>>1 mol% total impurities) may show retarded rates of transformation owing to the reduced probability of delivery of the "right" atomic species to the reaction site.

Impurity Effects on Phase Nucleation and Transformation

The effect of impurities on transformation is readily incorporated into a simple statistical treatment and becomes apparent particularly at transformations occurring below $T_h \approx 0.7$, with impurity contents greater than 5 to 10 mol%, as seen in Figure 4.9. These systems are sometimes called "stabilized" solid solutions and should not be treated as one-component systems. The noted effect on transformation temperature is as easily understood with reference to the freezing point depression of solutions as with the present statistical arguments.

Detectable and measurable impurities (minor components) are sometimes said to "chemically stabilize" a given structure type as their presence may affect the

---

[12]Defects in metallic materials will reduce heat transfer through increased phonon and electron scattering, while defects in ionic or insulating phases have much the opposite effect as the defects themselves are responsible for the vibrational and electronic modes necessary for efficient heat transfer.

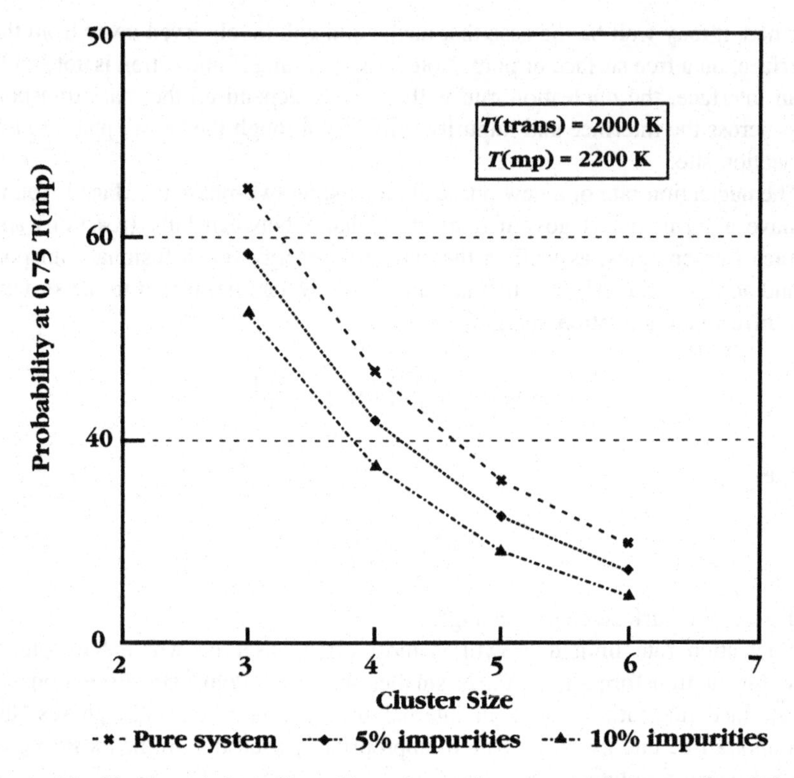

Figure 4.9 Simple statistical effect of impurity species on nucleation or continued phase growth. A greater retarding effect is seen with greater complexity (correlated probability) to form the nucleus cluster.

apparent equilibrium state achieved. The presence of impurities does alter the state of the system and may alter the energy balance between competing final state configurations if they are only slightly separated in energy in a pure system. A measurable impurity concentration is part of the full system description—it is possible that the altered system has a different equilibrium state than the pure system. Impurities whose solution results in compensatory lattice defects and increased diffusion rates may promote the growth of the stable phase; however, those same impurities will be a statistical hindrance to the nucleation of that phase. Impurities that remain in solution principally affect the atomistic path taken by a chemical system much more than the absolute outcome, although a colligative effect will be noted on the final state properties.

There is a situation in high molecular weight polymeric liquids or in inorganic multicomponent ($C > 4$)[13] systems where the complexities of the liquid and equi-

[13] $C > 6$ in metallurgical liquids.

librium solid phase do reduce the formation probability of a low-energy crystalline nucleus to near zero. In these systems, even low-viscosity liquids can be quenched to form glassy phases. The persistence of these phases and their stability against crystallization are the foundation of the glass and plastics industries.

Vapor–Solid Transformation

Vapor → solid transformation must be considered in oxide systems. The vapor–solid interface is not generally part of the "closed system" for consideration of phase equilibrium. At low temperatures particularly ($T_h < 0.5$), solid surfaces serve as efficient heterogeneous nucleation sites for a variety of vapor species that may be present in the process environment. At $T_h > 0.9$, however, most oxides show a detectable vapor pressure. At these high temperatures, condensation or crystallization from the vapor or involving vapor phase species may contribute in large part to equilibrium phase transformation. For some systems, chemical vapor transport is used for viable commercial scale growth of strain-free thin films and crystals.

Crystal growth from the vapor phase is limited principally by the rate of matter transport. A possible constraint on the attachment kinetics from the vapor phase may be the rate of charge transfer to and away from the surface, as molecular bonds in the gas phase must be broken and new bonds formed in the growing crystal phase. This is most often of concern in growth of highly purified insulating phases. Mobile electronic carriers in the solid can be highly efficient in transporting excess charge and heat from defective or impure surfaces.

As the delivery rate is slow at normal pressures, the latent heat of crystallization can usually be dissipated by radiation or conduction and will not be a limiting factor for growth. Heat transfer through "rebound" or kinetic transfer to gas molecules is also effective at moderate overpressures.

In saturated gas phases, dendritic growth from the available surfaces or "snowflake" type morphologies are often observed. These morphologies allow the maximum surface area for random attachment of atoms and for the radiative transfer of latent heat. The rate of growth will be seen to reflect a balance between the collision probability of atomic species from the gas phase onto the solid surface, a sticking probability (reflecting the generally small likelihood that bond formation will be favorable), and the rate at which heat can be transferred away from the interface. Anisotropic crystals are usually seen to grow with particular orientation relationships, representing the "easy" or most probable growth directions.

On large, flat surfaces, single crystal product layers may develop under ideal conditions, as in equilibrium phase growth by chemical vapor transport. For growth on a substrate, the largest crystal axis generally is parallel to the growth interface and the shortest crystal axis is found parallel to the growth direction. If the interface is not flat or if rapid transport paths exist (cracks or existing grain boundaries perpendicular to the gas–solid interface), the same principles hold for the polycrystalline layer that will develop.

For phase transformation via the vapor phase, the lattice match cannot be perfect between the initial solid (substrate) and growth product. Developing strains will influence the rate of growth and tend to lead to a polycrystalline product.

Phase Transformation in a Constrained System

For an equilibrium phase change to occur in a single phase (congruent) system, the system must be unconstrained by its surroundings and the surroundings must be capable of acting as a thermal reservoir in the traditional thermodynamic sense. Polymorphic phase transformation will involve some finite ΔH_{trans}, exothermic or endothermic, which must be either removed or supplied, respectively. Thus, an insulating surroundings, which cannot transport heat, will act to suppress an exothermic transition that occurs on cooling. Similarly, a phase that is constrained in an inelastic or incompressible matrix may not undergo a transition that results in $|\Delta V| > 0$ until the thermodynamic driving force exceeds the volume strain energy.

Excessive net volume change on solid-state transformation, or an extreme change in any unit cell dimension, will result in a net lattice strain. Strains can be minimized by defect formation in one or both phases, through heterogeneous nucleation such as at a free surface or internal pore, or through topotactic phase transformation (in which the new phase has a specific crystallographic relation to the old phase).

Anisotropic response to external fields may also result in volume strain. Piezoelectric devices such as transducers rely on the dimensional change with applied field to translate electrical energy into mechanical energy. However, all polarizable ionic lattices undergo some dimensional change owing to an imposed field and the resulting strain can influence phase transformation.

Susceptibility and magnetization in ferrimagnetic ceramics are generally anisotropic. Magnetic interaction can result in overall shape or volume change, described through the magnetostrictive coefficients of the material. Through magnetostriction, the second-order property can influence or be affected by first-order processes that affect or constrain molar volume. (See lamba-transitions.)

5

The Phase Rule and Heterogeneous Equilibria

5.1 The Gibbs Phase Rule

The phase rule relates the observed state of the heterogeneous system with number of thermodynamic variables required for the energetic description of its state as equilibrium. The phase rule is stated simply as the mnemonic, $\Phi = C - P + 2$, where Φ is the variance = number of state variables that must be independently fixed or determined, C is the number of components, and P is the number of phases observed. The factor of 2 arises from the normal number of intensive variables, temperature and pressure.

The phase rule is a guide to the interpretation of experimental observations and to the most appropriate system description. If two phases (P = 2) are observed in a one-component (C = 1) system, then $\Phi = 1$, meaning that the observation conditions (*T* and *P*) are not independent. Such a condition can only be observed at a single temperature if pressure is fixed. If two polymorphic phases are observed as coexisting in apparent time-independent equilibrium at several isobaric (same pressure) temperatures then, according to the phase rule, either the system is not at equilibrium (i.e., one or both of the polymorphs is thermodynamically stranded) or it is not a one-component system.

The phase rule is used particularly in the understanding of multicomponent, multiphase equilibrium. The one-component system is adequately described using the Clausius–Clapeyron equation for simple univariant or invariant phase equilibrium. The total energy at equilibrium is known simply from the values of the state variables mass, temperature, and pressure since coexisting phases have the same molar free energy.

The phase rule derived from a search by Gibbs for a functional form to relate the thermodynamic quantities of internal energy, enthalpy, and entropy to the physical description of a chemically heterogeneous, but not necessarily atomistic, system. The atomic nature of matter was not established for three decades after Gibbs published his work. The original description was of a macroscopic system that could be subdivided into hypothetical "cells," each being described as the smallest portion of the system for which the (macroscopic) state could be completely defined. Equilibrium is achieved on the unchanging coexistence of two or more adjacent cells. Modern microscopic methods have not altered the validity of Gibbs' work, but have altered the scale on which we may view the system.

Importantly, the phase rule does not rely on mathematical descriptions of the thermodynamic functions, simply requiring that there is a relationship between free energy, enthalpy, entropy, and composition for each phase. The phase rule sums up the available physical facts and compares that number to that needed for the simultaneous solution of the equations of total free energy at equilibrium of the system. The system is a fixed mass of known composition and has been isolated for study. No transfer of matter or energy is allowed across the system boundary: it is a closed system.

Statistical mechanics links the local atomic configurations and energy states to the global or macroscopic state of the system and readily extends the original Gibbsian concept of hypothetical system cells as well as establishing a basis for description and consideration of the random variations which are now known to characterize real materials. The meaningful scale of observation for phase equilibrium is still defined by the requirement that the state of each cell be fully determined through knowledge of the macroscopic state of the system. A system examined on the nanostructural scale may exhibit behaviors or phenomena that do not appear to follow the phase rule and may require the application of statistical methods in interpretation. It should be remembered particularly that the energetic realities of physical interfaces or surfaces are not considered in classical thermodynamics.

Derivation of the Phase Rule

An isothermal, isobaric, system having C components distributed among P phases has a total energy that is the sum of the partial molar free energies of each component in each phase, $P(C-1)$ composition variables. The component energies are dependent on the intensive variables temperature and pressure, so that to state the total energy of an arbitrary condition of the system, the values of $P(C-1)+2$ variables must be known.[1]

If the system is at equilibrium, the energy of the system is constrained by the condition that the partial molar free energy, or the chemical potential, of each component be the same in each phase present. This is the definition of phase

---

[1]As it is known that the presence of a magnetic field can influence the degeneracy of the stable electronic configuration of some transition metals, the phase rule may require modification to specifically include the magnetic flux as an intensive variable in some cases, causing the phase rule in such cases to be more correctly stated as $\Phi = C - P + 3$.

compatibility. The chemical potential is dependent on the intensive variables and the concentration of the other components in the phase. At equilibrium, there are then C chemical potentials and $(P - 1)$ distribution variables of the components among the phases present, or $C(P - 1)$ thermodynamic relationships which describe the system free energy.

Comparison of the number of variables required to state the system energy with the number of thermodynamic relationships available for their simultaneous determination results in the system variance, or degrees of freedom: $[P(C - 1) + 2 - C(P - 1)]$. Simplified, this difference is the "phase rule."

The phase rule is stated $\Phi = C - P + 2$ where Φ is the variance of the system, or the number of intensive variables that must be explicitly stated or determined to define the energetic state of the system at equilibrium. A system at equilibrium will "obey the phase rule," but apparent adherence to the phase rule does not guarantee that the system is observed at equilibrium.

Requirements for Calculated Phase Equilibrium

Each phase that may exist in a system has a free energy function dependent on phase composition, temperature, and pressure. Graphically, this function describes a multidimensional surface that must be continuous and must have continuous gradients in all dimensions. The free energy gradient for a phase subject to changes in applied pressure, temperature, or composition is given through the relation $\partial G = V\partial P - S\partial T + \Sigma\mu_i\partial n_i$.

The simple partial differential equations expressed in this relationship can be used to establish the thermodynamic basis for a phase stability diagram linking any two of the experimental variables. Complex partial derivatives may also be obtained with reference to Maxwell's equations, and can be of value in the interpretation of phase boundaries in complex systems.

Given a chemical system, it is possible to calculate the ideal energy functions of all possible phases, compare their gradients, and establish the conditions of phase equilibrium from first principles. Such ab initio calculation methods do have application where experimental observations are impossible such as at very high temperatures and pressures. More generally, energetic calculation is based on systematic chemical and structural variation of known crystalline prototype structures or ideal solutions.

The functional forms used to describe the compositional variation of free energy (the chemical potential) and entropy are somewhat arbitrary in mathematical terms but generally have their basis in the statistical description of mixtures described earlier. Continuum elastic theory can be used to describe simple pressure–temperature–volume relations. The mathematical forms generally describe the smooth variation in energy around a reference state or prototype as the phase composition, temperature, and pressure (or volume) are varied.

The reference free energy of the ideal or prototype phase is itself a relative value established through comparison with known reference states for the components. Multiphase equilibrium determination using mathematical methods requires that the reference state for the components be the same in the description of each phase. Experimental data for portions of the free energy function of each phase are determined using calorimetry, dilatometry, and other property measurements and used to "fine tune" the equation of state.

Simultaneous solution of equivalent chemical potential gradients using the calculated free energy functions for each of the equilibrium phases can be used to determine the effects of temperature or pressure on the compatible phase compositions. Such methods are generally used to extend limited experimental data in oxide systems, where kinetic limitations on equilibration severely restrict the number of reliable measurements of equilibrium compatibility and saturation surface curvatures.

Although ideally there is an equation of state that describes the phase behavior of the heterogeneous chemical system, regardless of its complexity, perhaps it is fortunate that the phase diagram description of the system does not require explicit knowledge of that relationship, but rather that such a relationship does exist and has a fixed number of variable quantities.

5.2 The Phase Diagram

While explicit calculation can be used to characterize behavior around a known system condition of composition, temperature, and pressure, a phase diagram compactly presents the equilibrium characteristics of a heterogeneous system under a broad range of conditions. For multicomponent equilibria, the phase diagram is the basic means of presentation of equilibrium physical information for a compositionally related series of systems.

Transitions occur in multicomponent systems that are not homogeneous, congruent processes. Incongruent processes include precipitation from saturated solution (liquid or solid solution), peritectic or peritectoid decomposition, phase formation, and eutectic or eutectoid reactions. A phase diagram is an effective tool for the presentation of the complex reaction processes between thermodynamically compatible phases in chemically complex systems.

Most oxide phase diagrams represent working hypotheses on equilibrium behavior, and are presented without significant mathematical detail. Through the diagram, a limited number of data points can be used to give a broad view of the equilibrium thermodynamics governing phase behavior and compatibility without recourse to mathematical description. Application of the phase rule in the preparation of a phase diagram ensures that the phase equilibrium described has an unambiguous thermodynamic description.

The diagram can usually be formulated and used on the basis of the graphical rules, and with the guidance of the phase rule, without any requirement for mathematical statement of thermodynamic relationships. No "rule" for presentation ensures that a phase diagram presents true equilibrium behavior; however, a valid phase diagram presentation is sure to belong to a small set of possible diagrams that includes the true equilibrium. Effective application of phase equilibrium diagrams becomes easier if the user is familiar with the varying forms that may comprise that set, as these will describe all possible stable and metastable equilibrium states that could be observed in practice.

The phase diagram is most often presented as a two-dimensional figure, although prismatic forms can be used and will be discussed briefly in a later section. A two-dimensional plot restricts the explicit presentation of information on phase equilibria to that which can be described through a dependence on a maximum of two independent variables. The phase rule is a means to evaluate the diagrammatic presentation of equilibrium, to guide the interpretation of the graphical elements in terms of multicomponent equilibria, and to assess the adequacy of the overall experimental description of the system.

Conventional presentation of an oxide system is as isobaric and unconstrained. Condensed oxide phases are not highly compressible and have low thermal expansion so it can be taken that the energy functions are not strongly dependent on external or mechanically applied pressure. It is common practice to reduce the number of explicit variables considered in the phase rule to reflect the assumed "constant pressure condition" and to apply the "condensed phase rule," $\Phi_c = C - P + \underline{1}$, in solid-state systems. Minor internal stresses due to reaction and phase development will not usually be sufficient to alter the course to phase equilibrium. At equilibrium, however, no internal pressure gradients may exist. A dense or constrained system cannot be considered isobaric unless sufficient consideration has been given to the relaxation of internal stresses.

Vapors are highly compressible, indicating a significant dependence of free energy on external pressure. If the application temperature or characteristics of the system require consideration of gas–solid or gas–liquid equilibria (often at $T_h \geq 0.9$), the effects of pressure on equilibrium may be significant. This could occur, for example, in the sealed containment of a system with potentially reactive vapors. An entrapped vapor phase is necessarily constrained and pressure must be explicitly considered in the application of the phase rule, as in final stage sintering where gases may be trapped in closed pores.

The phase diagram describes states of equilibrium that are observed only before or after a phase transition has taken place. It describes a system's state at constant pressure and temperature, and does not describe exactly the behavior of a real system subjected to gradients in temperature or mechanical stress or a combined stress state. As has been discussed, local gradients in chemical potential must exist for any transformation to go forward. The balance between local chemical potential gradients, the requirements of stoichiometry, and the capacity of the local structure to transmit thermal and mechanical energy will determine whether the transformation to "equilibrium" as portrayed in the phase diagram will proceed at a measurable pace.

The System Components

Phase equilibrium is presented in the phase diagram as dependent on temperature and system composition. The composition must be described appropriately in terms of the system components. The primary requirement for a working description of a multicomponent system is that the number of components is the same as the number of independent chemical potentials. The maximum number generally considered is the number of chemical elements, but this is often reduced by stoichiometric or structural arguments in dealing with condensed phase equilibria. There is little value in overdescribing the system; appropriate and thoughtful selection of stoichiometric or structural chemical groupings as components can aid in the physical understanding of the system's behavior.

The proof of an adequate system description is in the ability to determine the energetic state of the system unambiguously. Application of the phase rule to account for behavior near a phase transition or critical reaction equilibrium is impossible without a correct accounting of the number of components present.

Equilibrium coexistence of compatible phases requires that each component has the same chemical potential in all of the phases present. As the chemical potential is dependent on phase composition (including defects), it follows that each phase must have a composition range, even so-called line compounds, when observed in multiphase equilibria. All real materials contain solutes, "excess" chemical species (relative to the stoichiometric compound formula) and defects (relative to a perfect crystal lattice). Without a compositional range and the presence of defects, real processes in the approach to heterogeneous phase equilibrium cannot proceed.

The appropriate number of components can sometimes be greater than the number of chemical elements. While most compact oxides can be described simply with either elemental or simple stoichiometric components (depending on cation oxidation behavior, see below), most organic compounds are better described with reference to their identifiable functional groups and characteristic molecular configuration and conformation. The functional groups represent distinct and unique bonding configurations that reflect the nature of the underlying carbon chemistry. In the carbon–hydrogen system, it is appropriate to "overdescribe" the system's components, to include the functional groups such as $(C_6H_6)_n$, hexane, or methyl groups that characterize the equilibrium phases over a limited range of experimental pressure and temperature, as it results in a less ambiguous description than through simple bulk composition.

In complex oxides and mineral phases, structural elements can similarly be included in the system description using "virtual chemical potentials." The virtual potential energy functions are associated with the structural features of the lattice or coordination groups whose occurrence varies with overall chemistry, pressure, and temperature. These terms might describe the systematic incorporation of an interstitial solute and associated lattice distortion, for example, complex anions, or the structural units that make up ordered layer sequences in Magnelli-type phases. These quasichemical species are structural building blocks and, as in the carbon–hydrogen system, reflect significant features of electronic structure and bonding.[2]

The presentation of the system's components in terms of elements, stoichiometric, or structural units of varying complexity, does not affect the validity of a phase diagram or the application of the phase rule in its interpretation. However, there must exist an implicit correlation between basic chemistry and the concentration of the complex or virtual component species at the conditions of observation, such as from known equilibrium constants, and the virtual component units must be appropriate for the description of the behavior of the chemical elements in each and every phase present in the system under observation. For example, "SiO_2" is an appropriate unit for condensed phase studies below $T_h \approx 0.9$, but is not appropriate if vapor phase equilibria are considered as significant. "SiO" is then present. Similarly, the carbonate ion $(CO_3)^{2-}$ is a reasonable component for the description of condensed or aqueous mixed salt systems, but cannot be used as a component in the description of solid–vapor equilibria.

Structural and stoichiometric component descriptions link the phase diagram chemistry to the features of commonly encountered physical systems. The extension of such descriptions to extreme temperature and pressure conditions is done with caution, as the component description relies on a correlation function that may not have general applicability.

[2]The applicability of virtual potentials in the description of mineral phases is presented by Powell, Chapter 8 in Saxena (1983).

6

One- and Two-Component System Presentations

6.1 One-Component Systems

One-component systems undergo only congruent transformation or polymorphism. For these systems, $C = 1$ and the phase rule is stated $\Phi = 1 - P + 2 = 3 - P$. Logically, the variance cannot be greater than 2, and the presentation of phase equilibria in a two-dimensional temperature (T) versus pressure (P) plot is possible. The behavior of all individual chemical elements can be presented in one-component format, schematically given in Figure 6.1.

Polymorphism and condensed phase transformation of compounds may also be presented in this format if a consistent stoichiometric or formula unit is appropriate for all phases. Exact behavior during equilibrium transitions is indeterminate both mathematically and physically; complex components may undergo stepwise transformation across an interface during isobaric or isothermal transformation. The phase rule requires that only a single homogeneous phase may exist over a pressure–temperature range for the one-component description to be appropriate.

Any phase diagram is composed of graphical elements: divarient areas, univariant curves, and invariant points. In a one-component system, the maximum number of coexisting equilibrium phases is three at invariance, $\Phi = 0$. An area represents divariant equilibrium, $\Phi = 2$, and is the range in T–P space in which a single phase has the lowest free energy, but temperature and pressure must be known independently to know the free energy of the system. A univariant curve is the boundary between two single-phase areas and represents the condition $G_1 - G_2 = 0$; phases 1 and 2 have the same free energy along this curve and can coexist in equilibrium. For univariance, $\Phi = 1$, as two simultaneous equations exist to describe the free energy of the system. Either the known pressure or temperature may be used to unambiguously describe the state of the system.

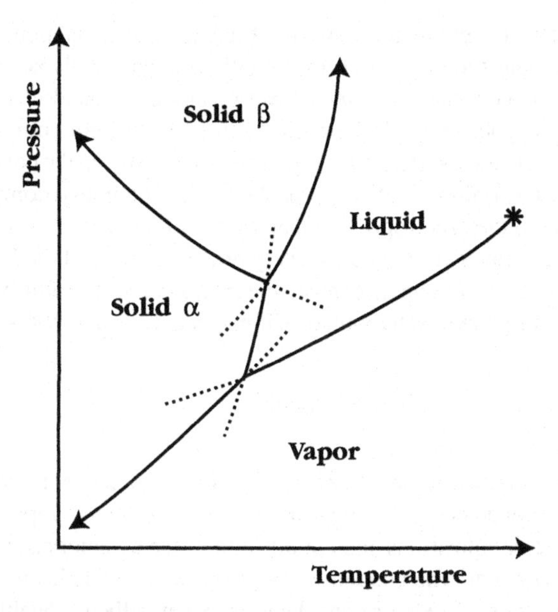

Figure 6.1 Schematic single-component system having two solid polymorphs, one liquid and one vapor phase stability region.

Experimentally, univariance or $\Phi = 1$ behavior implies that a temperature "arrest" will be observed for isobaric transformation from state $1 \rightarrow 2$. Univariant homogeneous transitions are commonly detected using differential thermal measurement, comparison of the internal temperature of the specimen with that of a reference material that is similarly heated or cooled. Small volume, packed powder samples are commonly used for solid-state studies. Observation of the true "thermal arrest" on transition is possible only if heat transfer through the experimental system is not limiting and if the particle size of the sample, while sufficiently small to ensure high and uniform surface nucleation rates, allows neglect of surface energy or defect "stabilization" effects. For melt solidification, accurate thermal measurement without contamination is often hindered by the need to measure through a crucible wall, which increases thermal lags in the system. As any thermal lags are experimental system size dependent, a systematic series of sample sizes may be investigated so that the true transition temperature may be determined.

For compound polymorphism or freezing of complex liquids, a transformation occurring over a temperature range ΔT that does not extrapolate to zero with decreasing sample size under isobaric conditions indicates that the system is not a one-component system.

For many systems in which experimental measurement of phase volume and transformation enthalpy exist, it is possible to construct simple one-component diagrams relating the known polymorphic phases in P–T space. This can be a purely hypothetical exercise or, if specific transitions such as melting points and normal boiling points are known, it may result in a reasonably accurate diagram presentation. The congruent polymorphism of any compound can be described using a one-component type *T–P* presentation if the system boundary excludes the vapor phase and the temperature range does not far exceed the condensed phase transitions. As the boiling point is approached, only compound compositions that also correspond to the azeotropic liquid composition in the truly multicomponent system can be described and then only a very restricted range of total pressure. Any *T–P* diagram for a compound that shows the liquid–vapor or solid–vapor equilibrium curves can be presumed hypothetical if presented as a one-component system.

6.2 Two-Component or Binary Systems

The most useful multicomponent system presentation is for isobaric two component (or binary) condensed phase equilibria. A temperature–composition plot can present all phase compositions and equilibrium phase assemblies unambiguously for this type of system. The condensed system usually excludes interaction with an equilibrium vapor phase from consideration, a generally reasonable assumption in refractory oxides observed near or below their melting points at normal pressures. A mechanical pressure of one atmosphere (0.1 MPa) is normally assumed. The properties of condensed oxides are not strong functions of applied pressures as generally used in processing; the phase diagram can be assumed appropriate for several orders of magnitude higher pressures (compression) or under states of moderate mechanical tension.

The isobaric condensed system is appropriate only under conditions that the equilibrium vapor pressures of the components are very small ($\approx$ 1 to 10 ppm) relative to the total overpressure. Experiments in vacuum or at temperatures where vaporization of one or both components is readily detectable may not yield results in accord with the phase diagram of the condensed system.

The two components may be single elements, simple stoichiometric groups, or complex chemical-structural groups (compounds). The selected components or "endmembers" must be congruently melting; otherwise the system is quasibinary or pseudobinary (see discussion below). The components must be appropriate to express the composition of each and every (condensed) phase that occurs. Vapor phase species are seldom described using the simple stoichiometries of the condensed phases. Equilibrium vaporization and condensation phenomena must be discussed with full consideration of elemental system equilibrium.

Equilibrium must be described in terms of the chemical potential gradient functions of the components as described, so a complex component concentration must have a unique correlation with the free energy function of each phase. The more complex the components selected, the more limited the range of application or extension of the phase equilibrium presented. Thus, although a "two-component" format can be used to present details of phase transition behavior in a complex solid solution, such as a four-oxide relaxor ferroelectric composition around the condition of saturation in a single oxide, it is not adequate to present a broad range of chemistries for that transition. The simple single oxide or stoichiometric oxide unit component is the usual component selected.

Equilibrium in a single or multicomponent system is defined by the condition $\Delta G = 0$, or that the overall free energy cannot be further reduced by a change in the phases present or the states of those phases. The additional requirement of multicomponent phase equilibria is that $\partial G_{T,P} = \Sigma \mu_i \partial n_i = 0$, which is to say that the chemical potential of each component in each phase is the same.

Phases that satisfy these equilibrium requirements are called "compatible." The chemical potentials are dependent on phase composition, so that compatible phases in a multiphase equilibrium have energetically fixed compositions at given values of experimental T and P. An equilibrium isothermal assembly of compatible phases will have minimum free energy and zero driving force for compositional change. The equilibrium proportions of the compatible phases for a given system of known composition are fixed by conservation of mass. This phase assembly yields the lowest total energy state of the system that is stable against statistical fluctuation and local atomic rearrangements.

Requiring equivalent chemical potentials of each component in each phase ensures that there is no driving force for an overall change in phase proportions at equilibrium, although a random transfer of species between phases as required by a statistical thermodynamic view of the system is not disallowed. Grain growth and other morphological changes in phase appearance or physical distribution such as "ostwald ripening" are possible and likely to occur over a long observation period. The system's state is unchanged by a change in the state of division of a phase, so long as the mass fraction of the system contained in that phase is unchanged.

Graphical Characteristics

The number of components or $C = 2$ in a binary system. The phase rule is stated $\Phi = 2 - P + 2 = 4 - P$. The isobaric "condensed" phase rule is commonly applied, $\Phi_c = 2 - P + 1$ or $\Phi_c = 3 - P$. As with the one-component diagram, a single phase area has a variance of 2, a boundary curve or line represents univariant equilibrium, and the intersection of univariant curves is at an invariant point, $\Phi_c = 0$. The characteristics of a binary diagram are shown by example in Figure 6.2.

All single-phase regions have some compositional width, although many are referred to as "line" compounds as the measurable compositional range is within the uncertainty of the experimental technique for characterization. The isothermal chemical potential of each component must vary smoothly across a single phase region.

The boundary curve of a single phase region indicates saturation and exsolution

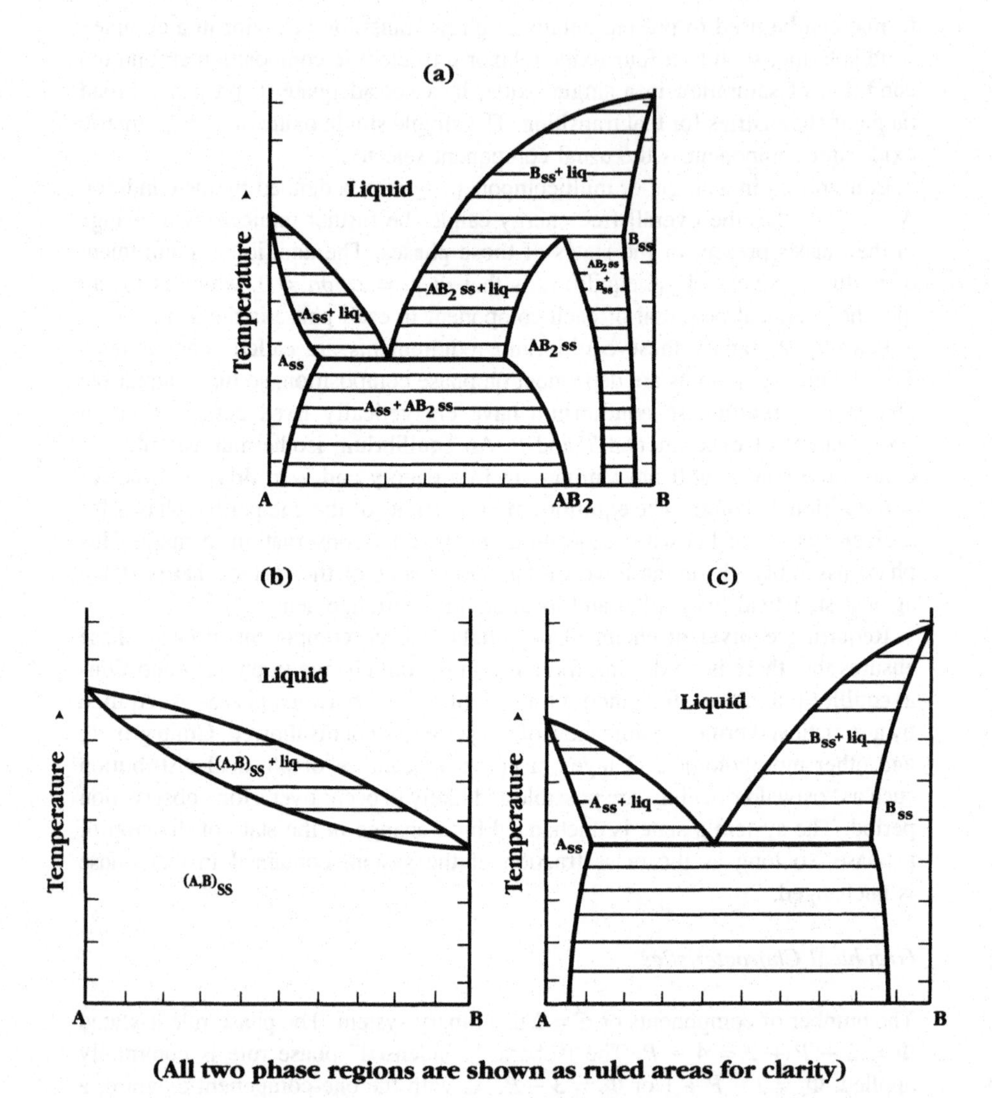

(All two phase regions are shown as ruled areas for clarity)

Figure 6.2 (a) Hypothetical system A–B with an incongruent compound AB_2. Ruled isotherms join compositions of compatible phases. Conventional graphical formats favor explicit labeling over ruling of two phase areas for compactness, with labels in-region if possible. (b) Hypothetical system A–B with complete solid and liquid solubilities. Comments as above apply on presentation. (c) Hypothetical system A–B with limit solid solubility and eutectic behavior. Note that the saturation surface of the liquid intersects the ruled isotherm between compatible solid solutions at the eutectic temperature defining a three phase invariance.

of a second phase, defining univariant equilibrium. The equilibrium compatible phase compositions must vary with temperature as the free energy functions vary with temperature and composition. Therefore, the univariant curve must have a notable slope in T–composition space and a curvature that relates to the dynamics of the saturated solution. For isobaric univariant two-phase equilibrium, a known T specifies the compatible phase compositions or knowledge of one of the phase compositions specifies both the temperature and the second phase exactly.

The low-temperature limit of single-phase liquid behavior is the liquidus, which describes univariant equilibrium in its regions of continuous curvature. Inflections on the liquidus correspond to a change in the nature of the univariant equilibrium described and are either eutectic, peritectic, or "monotectic" invariant reaction points. At the eutectic, the single-phase liquid solution is saturated with respect to two solids. A simple binary system contains only one eutectic, which is the liquidus minimum temperature.

A simple binary system can contain one or more incongruently melting compounds (congruent compound formation defines new systems; see "collected systems" in the next section). The peritectic reaction occurs at the incongruent melting temperature. On cooling through a peritectic, the solid in equilibrium with the liquid at high temperatures reacts with a portion of that liquid to form a new solid phase which is in equilibrium with the remaining liquid at lower temperatures. At a monotectic, found only where liquid–liquid immiscibility exists, the two liquids coexist at invariance with a solid phase.

The high temperature limit of a single-phase solid or solid solution with respect to liquid formation defines the solidus of that region. The solidus also defines univariant equilibrium in regions of its continuous curvature and can show inflections that correspond similarly to changes in the nature of the univariant equilibrium if, for instance, a polymorphic transition occurs in the solid solution phase. Although only the segments of the saturation curve that are associated with solid–liquid univariance are referred to as "the solidus," the saturation curve of the solid solution is continuous at all temperatures below the melting point. Characteristically, it will change curvature and direction at the temperature of the first liquid formation (eutectic or peritectic reaction).

A two-phase region is also called a "ruled area," ruled by the isothermal tie lines that connect compatible phase compositions on the saturation boundary curves. Each tie line is uniquely defined and describes the proportion and composition of the compatible phases in univariant equilibrium as dependent on temperature. In a two-phase region involving two "line" compounds of a binary system, the tie lines are not uniquely defined by phase composition, only by temperature. The remaining uncertainty in phase composition and component activity requires that the total system variance is actually greater than one. The phase proportions at temperature can be determined using the lever rule, but the saturation boundaries are insufficiently defined for any relationship between composition, temperature, and chemical potential to be inferred.

The solidus and liquidus are often experimentally determined through the exploitation of the thermal characteristics of univariant and invariant equilibria. Invariance can occur only at a fixed temperature that will persist until one or more phases disappear (i.e., melt or crystallize). Controlled rate cooling or "free cooling" of the homogeneous liquid phase[1] may be used. For free cooling, the onset of univariant equilibrium is noted as a change in the sample cooling rate which reflects the enthalpy of crystallization. A temperature "arrest" denotes the occurrence of an invariant equilibrium, generally peritectic or eutectic. Each further temperature "arrest" denotes an additional invariant equilibrium reaction characteristic of the sample composition.

Many oxide systems must be investigated in the solid state or with only a limited amount of liquid present. Various properties may be used to detect the presence of melting or other phase change. The mechanical stability of a shape, as a pyrometric cone, may be used with deformation noted when the presence of several percent liquid allows plastic flow. The flatness and optical reflectivity of a polished surface also changes dramatically at melting. Changes in thermal conductivity and electrical and magnetic properties can be used to detect phase change and liquid formation in some well-characterized systems.

The homogeneous solid state is, however, more difficult to achieve as an experimental starting point than the homogeneous liquid. Temperature measurement and control also become increasingly problematic at high solidus temperatures.[2] Uncertainties of ±20 to 40°C are commonly reported in the eutectic determination of refractory systems whereas uncertainties of only <±5 to 20°C are normal in systems where homogeneous melts below 1000°C can be obtained.

Collected, Quasibinary, and Other Presentations

If one or more congruently melting compounds can form from the simple endmembers, the diagram is a collection of $x + 1$ true binaries, where x is the number of congruently melting compounds in the system. Incongruently melting compounds cannot be considered as endmembers if the system is considered at any temperatures above their decomposition or melting temperature.

To the ceramist, the conventional use of simple oxide endmembers and combined presentation of all true binaries that are chemically related through simple

[1]Similar to DTA (Differential Thermal Analysis) or DSC (Differential Scanning Calorimetry), the controlled cooling rate is maintained through external resistance heating. The power required to maintain the constant rate in the crystallizing specimen can be compared to that required for an inert reference and differentially related to the enthalpy of crystallization.

[2]Routine temperature measurement to 1600°C (2000°C in reducing atmosphere) is reasonably accurate using thermocouple techniques and optical pyrometry can extend the range of measurement to 3000°C. At these temperatures, however, direct observation of the specimen in a reliable noncontaminating containment is difficult to achieve. The use of indirect techniques or observation of quenched specimens adds to the uncertainty in refractory phase equilibria.

ratios of these endmembers is useful. It is common to use simple oxide raw materials for reaction and equally common that the first product of reaction will be selected by kinetic and not thermodynamic factors. Ceramic reactions often involve subsolidus phase formation and decomposition, so this convention is also highly practical for the understanding of the system and the stable and metastable relationships between phases.

A quasibinary can be defined as a multicomponent system for which a simple binary representation is appropriate, at least over a restricted range of experimental conditions. The diagram adheres to the phase rule and phase proportions can be calculated from its geometry. The phase relations of a quasibinary system may "not obey the phase rule" outside of these restricted conditions of experimental P and T. For example, a binary-type presentation involving the oxides of transition metals may have an unstated dependence on oxygen activity that is not significant for phase equilibria at low temperatures. Under a broad range of conditions, the system needs consideration of the true nature and number of its components and the simple "binary" is seen as a quasibinary section of the ternary system.

Quasibinary presentations also include systems in which incongruently melting compounds are used as endmembers for a system presented at temperatures below the incongruent melting point. The significant feature of quasibinary presentation is that the endmembers as stated are appropriate as components in the calculation of energetic functions over the restricted range of conditions presented.

Most oxide systems are presented with simple oxide endmembers. A simple system of two oxides should truly be considered as quasibinary as it is a subsystem of the true ternary "metal 1 + metal 2 + oxygen." Stated stoichiometric relationships defining the endmember oxides are equivalent to the selection of a quasibinary vertical section in the true ternary. Such a section should be termed quasibinary if the author or other presenter is aware of nonstoichiometric behavior in either endmember, otherwise the conventional reference to a "binary section" is adequate.

A pseudobinary presentation, on the other hand, is not sufficient for the inference or estimation of energetic functions or thermodynamic relations between the phases under the full range of conditions presented using the stated endmembers as components. Pseudobinaries may contain limited regions of quasibinary behavior, most often in the subsolidus regions, but are usually characterized by at least one incongruently melting endmember or phase(s) that are otherwise unstable prior to melting. Pseudobinary sections or "cuts" are commonly investigated in the preparation of true ternaries, particularly the liquidus and solidus. Pseudobinary sections of complex ternary or higher systems are commonly prepared as a graphical convenience to describe industrial processes.

Functional similarities among elements and their compounds can be used to derive effective presentations of phase relationships in related systems. It is reasonable to use the periodic similarities in elemental behavior in grouping similar systems and preparation of general composite phase diagrams, but the information conveyed can never be adequate to fully describe the thermodynamic state of a specific system that belongs to the family of systems described, nor can the composite presentation be termed a phase equilibrium diagram.

For example, it could be accepted that R_2O_3–MO_2 systems where R is a lanthanide rare earth metal element (all having similar ionic size and charge characteristics) and M is a either Zr or Hf (both group IVb cations of similar size) will be likely to have similar compound stoichiometries and also be likely to exhibit nontrivial degrees of solid solution. A schematic phase diagram of the "system" $R(=La + Nd)_2O_3$–$M(=Hf + Zr)O_2$ could be presented to graphically indicate phase stoichiometry common to all four quasibinary systems. The schematic cannot be considered as a phase equilibrium diagram unless presented with additional compositional information which would enable its positioning in the quasiquaternary system, but it does convey important information on the phase behavior of a class of ceramic materials in a highly compact manner. Similar groupings of elements are possible throughout oxide systems when functional substitution of one element for another owing to similarities in size and charge is reasonable to assume.

Equilibrium in an Open System

Application of the phase rule requires that the system be closed; mass is conserved. Condensed phase equilibrium further assumes that no significant involvement with the vapor phase, or gas–solid equilibrium, occurs. Although these requirements can be satisfied in many oxide systems that are not physically isolated from their environments, many useful applications of phase equilibria involve nontrivial interaction of a mobile species, a component of both the condensed system and its surroundings. Examples are found in nonstoichiometric oxide sensor and battery materials, "consumable" oxide refractory systems, bonded refractories, and metal–metal oxide systems.

For significant effect of a mobile species on phase equilibria, the species must be completely free[3] to move in the condensed phases within the time of observation and must be able to leave the vicinity of the interface.

The phase rule can be applied to describe phase equilibrium only in the case of exchange of a mobile species with a fixed volume of vapor phase—an effectively closed system. The system for observation, although it contains the vapor phase and exchange of species with the vapor affects the condensed phase equilibria, is isolated from the larger surroundings. The full phase rule may be applied for the binary system, which is appropriately described using the elements as components. $\Phi = C - P + 1$ (1 atm mechanical pressure assumed), so that the

[3] A reasonable mobility in the condensed solid phase would allow diffusional transport on the scale of the particle size in the time of observation. For a 1-μm particle size powder allowed to equilibrate for 10^4 s (1 week), $D \approx 10^{-12}$ cm^2/s is sufficient; for a 1-cm slab, $D \approx 10^{-8}$ cm^2/s is required.

system is invariant at $P = 3$, two condensed phases and a vapor phase, having a fixed activity of the mobile species at a fixed temperature.

If a species of the condensed phase can freely exchange and be mixed via free or forced convection into an infinite surroundings, the system is open. In an open system, the phase rule does not apply and the equilibrium state is indeterminate, although a steady state may be obtained over long periods of time owing to boundary layer formation. This situation is found in a freely vaporizing system, or in cases of dissolution into an infinite (far from saturated) solution.

If the species is common to both the condensed phase or phases and its surroundings, and the surroundings are of sufficient volume so that they can act as an infinite source/sink for that species, the system will have an equilibrium that is described through comparison of the formation energies of the condensed phases that may occur, given the temperature of observation and the fixed activity of the mobile species. Generally only a single condensed phase will be observed at equilibrium in this case.

Open system conditions are sometimes "overlaid" on the phase equilibrium diagram of the closed system. An example is given in Figure 6.3: 6.3(a) represents the invariant three-phase equilibrium MnO + Mn_3O_4 + gas (assuming that the system is at 1 atm (0.1 MPa) mechanical pressure, so that the phase rule, $F = C - P + 1$ applies). Figure 6.3(b) is the binary condensed system overlaid with oxygen isobars. Note that the true scale of diagram (b) is the oxygen/metal ratio, not uncommonly seen in this type of presentation. Without the isobaric information, the oxygen/metal ratio defines the composition of a closed system that excludes the vapor phase. In the indicated two-phase regions, oxygen and metal atoms are distributed only between the condensed phases.

With the isobaric information, Figure 6.3(b) can be interpreted in the open system: for example, at $p_{O_2} = 10^{-2}$ atm, Mn is oxidized completely to Mn_3O_4 at all temperatures below the transition at ≈ 1430°C. At higher temperatures the equilibrium solid is MnO_{ss}; melting occurs at ≈1670°C. Similarly, at lower partial pressures of oxygen, single-phase Mn_3O_4 is seen as the stable solid to progressively lower temperatures while the melting point of the reduced MnO_{ss} generally increases. The indicated equilibrium transitions are consistent between the two presentations. The equilibrium MnO + liquid (+ O_2) is better shown in Figure 6.3(b), although it has been indicated in (a) for comparison.

(At greatly reduced oxygen pressures, MnO vaporizes significantly above 1200°C and description of the equilibrium with the vapor phase through its oxygen partial pressure is no longer adequate.)

Most commonly, the equilibrium of multivalent metal oxide systems is used to fix the "reference" oxygen activity for sensing devices employing oxygen electrolytes or to fix a local value of oxygen activity during sintering or annealing, as a pellet or component within a packed powder bed. Low oxygen potentials are desired in metal melting operations and can be effectively controlled using composite oxide/carbon refractories. In bonded refractories, such as "tar" (carbon)

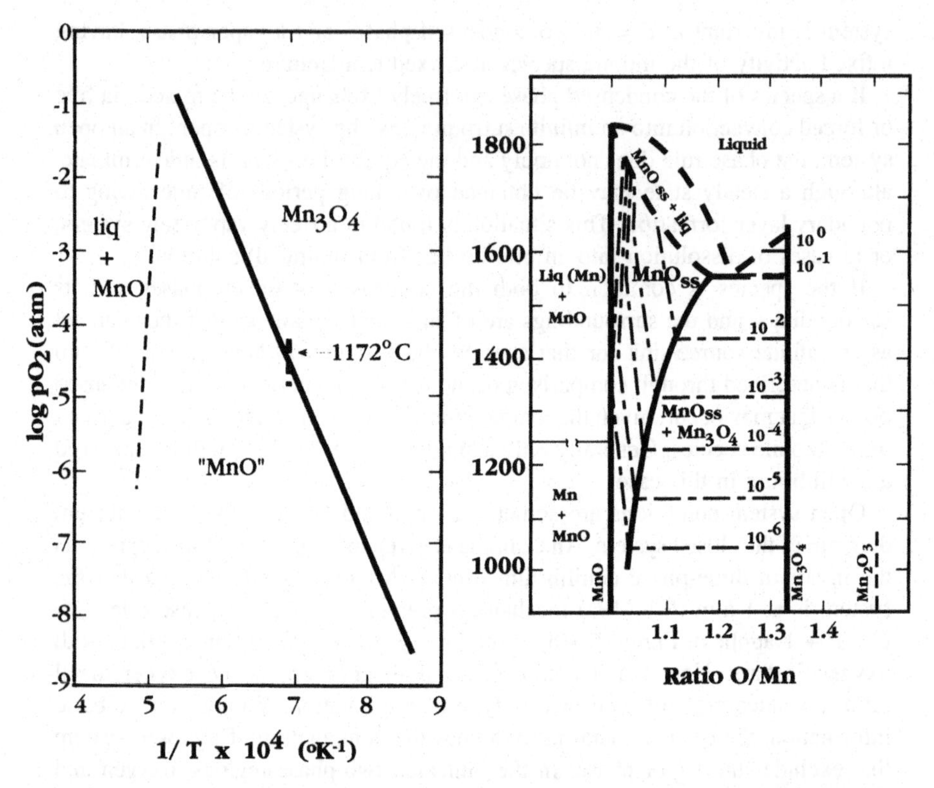

Figure 6.3 (a) Univariant boundary MnO–Mn_3O_4 [adapted from K. Schwerdtfeger and A. Maun, *Trans AIME*, 239:1115 (1967)]; and (b) MnO–Mn_3O_4 condensed system presentation with oxygen isobars [adapted from A. Z. Hed and D. S. Tannhauser, *J. Electrochem. Soc.*, 114:316 (1967)].

bonded MgO or graphite/MgO refractories, oxygen activities in a steel melt are controlled through the primary reaction $C_{(refractory)} + 1/2O_{2\,(dissolved\ in\ metal)} = CO_{(g)}$ or through the secondary reaction $MgO = MgO_{1-x} + (1-x)1/2O_2 \Rightarrow 1/2\ O_2 + C = CO_{(g)}$ (the latter being a three-component equilibrium).

At high temperatures and under extreme conditions, many "simple" binary systems break down through direct interaction with the vapor phase or interaction with corrosive liquids. Extreme examples are found in glass tank and crown refractories where mobile species from the melt and combustion vapors interact with the condensed refractory phases. (These systems and applications are further discussed in Chapter 8.) Alkali species are particularly mobile in the melt/refractory interface; alkali, sulfurous, and halide vapor species are all present in the corrosive vapor phase. The mobile species saturate the refractory over time, leading to phase alteration. Glass or metal contact refractories are considered "consumable" where their normal application anticipates a physical interaction that is not harmful to the product contained.

The refractory application environments are not generally isothermal, so steady-state concentration gradients may develop that are maintained owing to unequal diffusivities at the hot and cold faces. In the steady state, condensed phase development is locally consistent with the requirements of phase equilibrium of the local system. The stability of the refractory system is often dependent on the steep thermal gradient; true isothermal equilibrium in the open system will result in complete destruction of the refractory over time. For refractory containment systems, particularly melt containment, a steep thermal gradient ensures an extra margin of safety against melt-through and may be a good economic and safety tradeoff against extra insulation to reduce waste heat.

The Vapor Phase in a Closed System

Binary phase equilibrium in a closed system involving one mobile component and the vapor phase is commonly presented in the form of an Ellingham diagram. The diagram presents invariant equilibrium of the type $M + X_2$ (gas) $= MX_2$, where M and MX_2 are condensed phases. For the reaction at standard state conditions, $\Delta G^\circ = \Delta H^\circ - T\Delta S^\circ = -RT \ln K_{eq} = RT \ln(p_{X_2})$. Comparison of the first and last statements of the equality reveal that a plot of $\Delta G^\circ = RT \ln(p_{X_2})$ may be described by a straight line on a plot of ΔG versus T, with intercept of ΔH° and slope $-\Delta S^\circ$. Ellingham (1944) observed that experimental data for oxidation and sulfidation of many metals was well described through this simple representation. The line of invariance is continuous through a condensed phase change of the reactant or product, although the slope and intercept of the line do change abruptly at that point of transition. The equilibrium vapor pressure at any temperature is determined by numerical solution for p_{X_2}.

$\Delta G^\circ = RT \ln P$ is also the statement of free energy change of an ideal gas undergoing a change in pressure from $P_1 \rightarrow P_2$ where P_1 is the standard state, 1 atm. In practical Ellingham representations, a secondary right-hand nomographic scale (first suggested by Richardson (1948)) is added so that the equilibrium pressure of any reaction, normalized for 1 mol of X_2 gas reactant, can be determined at any temperature of interest. A line, constructed from ($\Delta G = 0$, $T = 0$ K) through the reaction line at the temperature of interest will intersect the nomographic scale at the equilibrium gas reactant pressure.

A generalized diagram for the reaction of many metal species with the same gas reactant is generally prepared. On such a generalized diagram, the intersection of reaction lines for dissimilar metals indicates the invariant characteristics of that three-component system for the invariance of the type $M + RX_2 = R + MX_2$ (T, p_{X_2}). It should be noted that the presentation of a reaction of the type M(s) + X_2(g) = MX_2(g) can be presented on the Ellingham diagram, but represents a univariant equilibrium (two components and two phases, condensed phase + vapor). The nomographic scale represents the p_{X_2} obtained for $p_{MX_2} = 1$ atm, although a numerical solution of the equilibrium constant can be obtained for any fixed value of product partial pressure.

A simplified Ellingham diagram for the oxidation of common metals is presented in Figure 6.4. In the commonly used form as shown, the several scales presented along with the simple nomographic oxygen scales represent the required

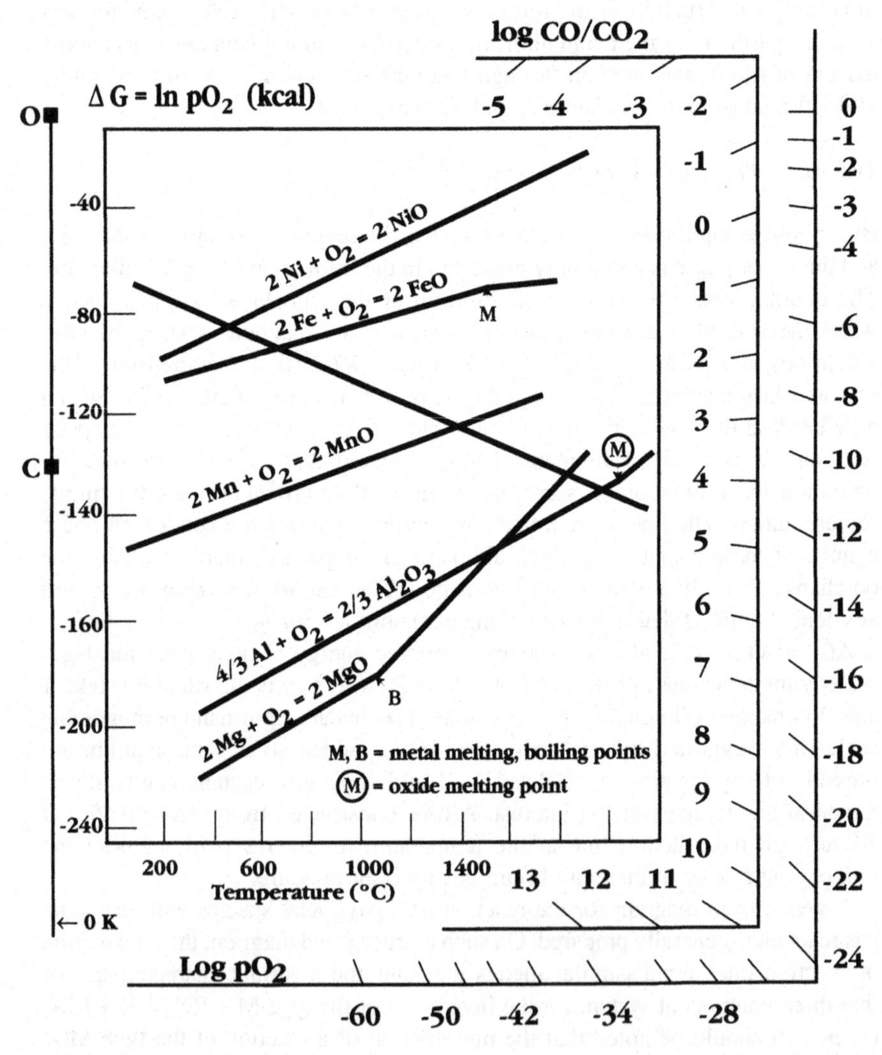

Figure 6.4 Simplified Ellingham diagram for Ni, Fe, Mn, Al, Mg, and C, with nomographic scales for p_{O_2} and CO/CO_2 ratio.

experimental mixed gas atmospheres to impose the equivalent p_{O_2}. For the bonded refractory discussed above, the presence of carbon is seen to be sufficient to reduce iron oxides in the molten metal. Similarly effective (and equivalent, thermodynamically) is an imposed atmosphere having a fixed $CO/CO_2 > 10^{3-4}$ ($p_{O_2} < 10^{-8}$, 1600°C).

For traditional ceramists, these equilibria are significant in sintering of fine ceramics and development of glaze colors, where uncontrolled reduction of metallic impurities in the combustion atmosphere can cause staining of whiteware or uneven coloration. In electronic ceramics, cofiring of metallic contacts requires consideration of the metal(1)–metal(2) oxide–O_2 equilibrium conditions, of particular concern for ceramic compositions involving the transition metals which are readily reduced.

The Ellingham diagram describes classic oxidation equilibria in a metal–oxygen system. Oxidation is an electrochemical reaction, requiring the ionization of the metal prior to bond formation with oxygen. Any reaction requiring electron transfer can be described in terms of the electrical potential required, ϕ, as derived using the Nernst Equation

$$\phi^{o} = \frac{\Delta G^{o}(T)}{nF}$$

where

n = number of electrons transferred
F = Faraday constant

and the superscript "o" indicates the reactants at unit activity. The effect of concentration or oxygen activity can be included in the standard manner,

$$\phi = \phi^{o} - (RT/nF) \ln K_{eq}$$

The reduction potential of an oxide is readily calculated using the information from the Ellingham diagram at the temperature of interest. For example, the Ellingham diagram indicates that Al_2O_3 would be reduced to metal at 1000°C if the p_{O_2} were below 10^{-34}. The reaction is written (for the Ellingham standard representation) as $4/3\ Al + O_2 = 2/3\ Al_2O_3$, requiring the transfer of 4 electrons for the oxidation of the metal, with $\Delta G^{o}(1000°C) = -202$ kcal/mol. Using the Nernst equation, the reduction potential is calculated as $\phi^{o} = -0.58$ V. Application of this bias voltage would reduce the oxide to metal at 1000°C and $p_{O_2} = 1$ atm. At reduced partial pressure, the reduction potential is similarly reduced, to −0.41 V at $p_{O_2} = 10^{-10}$ atm.

The Ellingham diagram information enables straightforward analysis of the equilibrium behavior of oxide systems at high temperatures and their tendency toward nonstoichiometric behaviors. Electrical potential gradients are present in many heterogeneous systems, and may be sufficient to initiate local redox reaction, resulting in oxide nonstoichiometry or phase decomposition.

The imposition of a controlled bias potential to control oxidation reaction of metals is termed "cathodic protection." In metal–water systems at or near ambient temperature, it is customary to present similar oxidation equilibria in terms of an isothermal Pourbaix diagram. The Pourbaix diagram is a representation of thermodynamic equilibrium phases in terms of the intensive variables and so is truly a phase diagram. The general diagram of this type is a plot of solid phase stability in respect to aqueous solution pH and electrical potential. The system is a three-component one: metal, hydrogen, and oxygen. The pH ($= -\log[H^+]_{aq}$) is a measure of hydrogen activity, or μ_H. Pourbaix diagrams are most often used in studies of aqueous corrosion of metals; however, the stability of oxide, hydrates, and hydroxide phases are sometimes similarly presented.

The use of Pourbaix diagrams in the investigation and prevention of metallic corrosion is outside the present discussion of oxide phase equilibria and is covered in many excellent texts on electrochemistry.

The Ellingham diagram information for oxides of multivarient cations is sometimes plotted as a Pourbaix–Ellingham diagram, where the axial presentation is of p_{O_2} versus reciprocal absolute temperature. Pourbaix–Ellingham diagrams are used to present metal oxide stability data and also "electrolytic domain" extent in oxide electrolytes.

6.3 Binary Phase Compatibility and Mass Balance

The binary diagram's general characteristics are related to the underlying nature of the condensed solution phases, their relative ideality, and component identity. The characteristic curvatures, the one- and two-phase regions, are topological features of the minimum free energy–composition surface as projected onto the *T-X* plane. The scaling of that plane, particularly the composition or "*X*"-axis, does not affect the topology but can affect the ready interpretation of the diagram and its application to experimental observations using the "lever rule."

The lever rule is a simple mathematical device to apportion the components in the equilibrium phases while conserving mass in the closed system. The lever rule is applied on an isothermal tie line in a two-phase or "ruled" area of a binary diagram. The calculation is explicitly performed for a single system composition at a known isobaric temperature.

An *isopleth* is a line of constant system composition; an *isoplethal study* is a detailed examination of the equilibrium phase assemblies and their proportions performed for a composition of particular interest over a range of temperature. The phase proportions so calculated will be stated in terms of the endmembers of the diagram and the diagram itself may be scaled in terms of concentration of the endmembers in either wt% (or fraction), mol%, or (less commonly) "equivalent" (equiv.) %.

Scaling the Diagram

If a diagram is scaled for mol% or mol fraction of components, the calculated phase proportions will be expressed in terms of the mol fraction of the system that is present in each phase. These fractions do not intuitively relate to the physical makeup or microstructural characteristics of the system. For this reason, many diagrams are expressed in terms of wt% of the components, as calculations of phase proportions are then expressed directly in terms of the relative masses of the compatible phases and can be directly related to the results of quantitative analysis.

Calculation of phase proportions in the equilibrium assembly can be performed correctly only if the diagram scaling information is clearly stated. Figure 6.5 demonstrates an extreme example; a mol of Li_2O weighs 30 g and a mol of Ta_2O_5 weighs 442 g. Three compounds form: $LiTaO_3$ (1:1), $LiTa_3O_8$ (1:3), and Li_3TaO_4 (3:1). These correspond to 6.4, 2.2, and 16.9 wt%, or 50, 25, and 75 mol% Li_2O, respectively. A 50:50 weight mixture of these components would yield an equilibrium product containing 37.5 wt% Ta_2O_5, with the balance being Li_3TaO_4. A 50:50 molar mixture would result in single-phase $LiTaO_3$.

Cation or anion equivalence is sometimes used to "normalize" component formulae. In a "cation equivalent" statement, each mol of each component as stated supplies the same number of mols of cations. A 50:50 cation equivalent mixture (1/2 Li_2O:1/2Ta_2O_5) would in this case be the same as a molar mixture. However, if the system Li_2O–WO_3 had been considered, a 50:50 cation equivalent mixture (1/2Li_2O:WO_3) would correspond to a 33:67 molar mixture of Li_2O and WO_3. If anion equivalence were used for defining scale, a 50:50 anion equivalence mixture of Li_2O and Ta_2O_5 (Li_2O:1/5Ta_2O_5) would correspond to a 83:17 molar mixture (about 25 wt% Li_2O).

Weight percentages are useful scaling when batching considerations prevail; any lever rule calculation performed using a "wt%" diagram will give weight percents of the phases directly without further conversion. Molar percentages are more relevant to crystal chemical studies and comparisons between systems having cations with similar periodic properties, while cation equivalences are useful in studies of substitution or acid–base type reactions. Anion equivalence is the most relevant if the phase diagram is to be used to interpret volume percentages or if polished sections are used to determine the equilibrium phases and proportions, as most oxide molar volumes correspond fairly closely to equivalent close-packed oxygen lattice volumes, irrespective of cation mass or charge.[4]

---

[4]A mol of close-packed oxygen atoms occupy about 10 cm^3. It is important to note that the most common oxides on earth, SiO_2 and H_2O, do not have close-packed oxygen lattices in the solid state at normal pressures. Both exhibit a diamond structure and show significant expansion on freezing. This "anomalous" behavior probably accounts for the persistence of life on earth, as ice floats on water, moderating the ocean temperatures, and silica-based materials (70% of the earths crust) expand on crystallization from the melt, giving a rather strong and solid surface to the relatively molten earth.

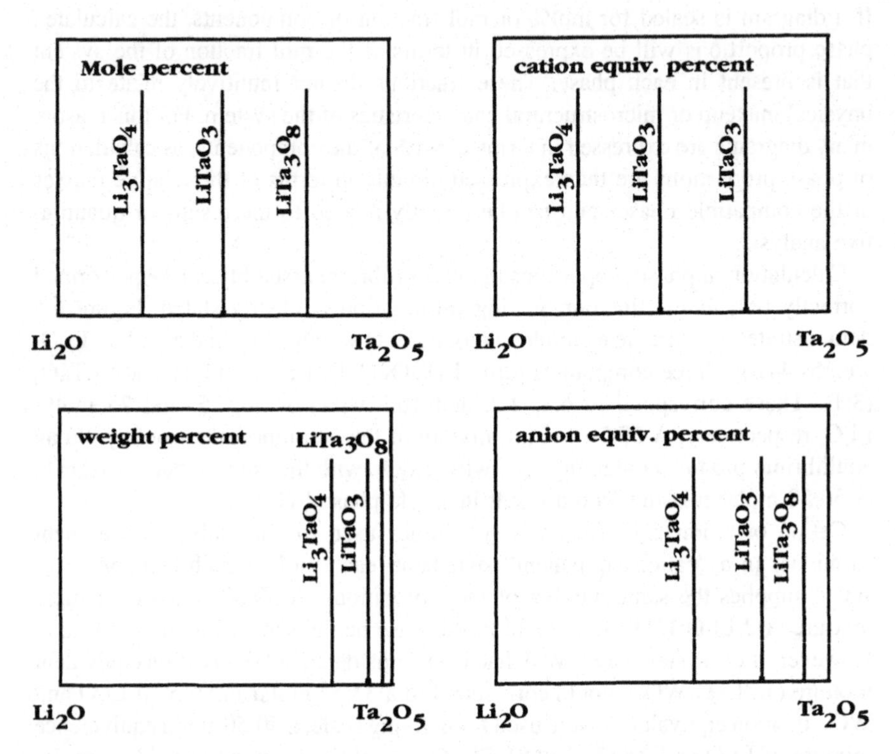

Figure 6.5 Example for calculation of phase proportions using the lever rule as applied to varying diagram presentation of relative mass or equivalence of components.

Li_2O	30 g/mol	2 cation equiv/mol	1 anion equiv/mol
Ta_2O_5	442 g/mol	2 cation equiv/mol	5 anion equiv/mol

	Relative position		
Diagram scaling	*$LiTaO_3$*	*$LiTa_3O_8$*	*Li_3TaO_4*
mol% Li_2O	50	25	75
mol% Ta_2O_5	50	75	25
cation equiv% Li_2O	50	25	75
cation equiv% Ta_2O_5	50	75	25
anion equiv% Li_2O	16.67	6.25	37.5
anion equiv% Ta_2O_5	83.33	93.75	62.50
wt% Li_2O	6.35	2.21	16.92
wt% Ta_2O_5	93.64	97.79	83.08

A diagram may be rescaled for the convenience of the user and current application. For rescaling a diagram, each critical point and compound composition should be recalculated individually to avoid serious error. No topological change to a diagram should result from correct rescaling. Comparisons between chemical systems should always be made using phase diagrams that are consistently scaled.

Crystallization Analysis

The process of determination of the phase proportions in a ruled area is aptly named "application of the lever rule," in reference to the mechanical analog of balancing a lever with a fixed fulcrum. In our analog, the position in space of the two ends and the fulcrum point are stated with reference to the composition scale of the diagram.

The system composition or isopleth is known, so the system is energetically determined at equilibrium if either the temperature or the chemical potential of one component is known also. It will be recalled that the isobaric chemical potential of a component is uniquely related to the temperature and phase composition, so generally it is said that the "composition" of one phase or the temperature must be stated. As solid-state solution limits can be smaller than the limit of error of chemical analysis for so-called "line compounds," it is practical to require that it is the temperature that determines the state of a solid–solid equilibrium system. A liquid–solid equilibrium can as readily be uniquely specified through knowledge of the liquid composition.

The endpoint phase compositions of the isothermal tie line must be stated in the same manner as the system composition. For example, a system of composition X_s is in equilibrium at T_1. The phase diagram indicates that the isopleth of composition X_s intersects the isotherm T_1 within the two phase ruled area "liquid + β_{ss}." The endpoints of the tie line are $X_{L(T_1)}$ and $X_{\beta(T_1)}$. These endpoints are the compositions of the compatible phases in equilibrium at T_1 for all systems whose overall compositions are in the range $X_{L(T_1)} < X_s < X_{\beta(T_1)}$. The relative proportions of the two phases that will be present in the equilibrium system are determined by conservation of mass.

Conservation of mass requires solution of two simultaneous equations (for composition stated in terms of percent B). The situation is graphically presented in Figure 6.6.

Returning to the lever analogy, it will be noted that the magnitude of the denominator of each fraction is the lever length (L); $L = X_{\beta(T_1)} - X_{L(T_1)}$, and the numerator is the same magnitude as a simple line segment length taken from the fulcrum at X_s to the end opposite the mass at $X_{L(T_1)}$; $X_s - X_{\beta(T_1)}$.

With reference to a composition-free energy diagram for this system at T_1, Figure 6.7, it will be seen that this distribution of the components yields the lowest possible total energy for the system. A single-phase solid or liquid, or a mechanical mixture of any other solid or liquid compositions that could be

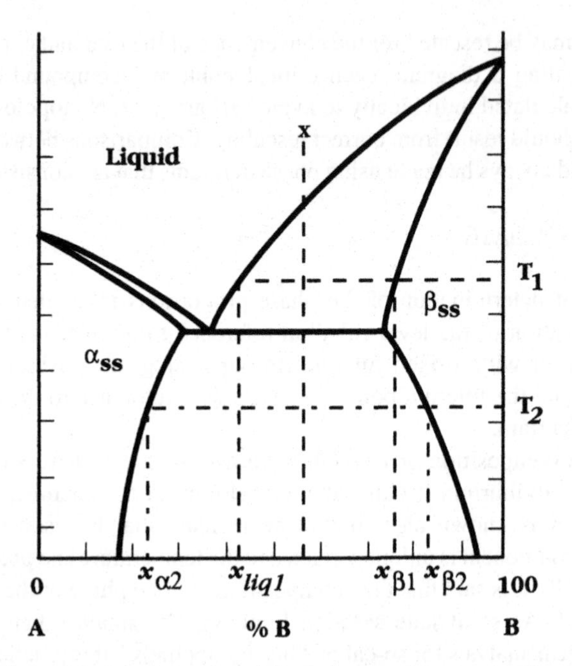

Figure 6.6 A lever rule calculation is valid only in two-phase areas. At each temperature, each true binary has unique lever length. Measurements are in arbitrary units. The system has overall composition "x". At T_1, $L = | x_\beta(T_1) - x_{liq}(T_1)|$

$$\text{fraction of liquid of composition } x_{liq\,1} = \frac{|x_\beta - x|}{L}$$

$$\text{fraction of liquid of composition } x_{\beta 1} = \frac{|x - x_\ell|}{L}$$

prepared within the constraints of mass conservation, would have a higher total free energy and also would not satisfy the requirement of equivalent chemical potentials of each component in each phase. (note that solids α and β could coexist and satisfy the conservation of mass and equivalent chemical potential requirements, but in a *true metastable equilibrium*).

A range of compositions that exhibit the same phases at every temperature share the same crystallization type. The proportions of those phases would vary for each composition within one crystallization type, as determined through conservation of mass of the components. It is readily apparent on examining many two- (and three-) component diagrams that although an infinite number of unique compositions can be described in any system, the types of equilibrium phase assemblies obtained are limited to very few.

Crystallization in a binary can be generally simplified through the division of the diagram into vertical composition bands, each vertical line intersecting a triple point or corresponding to a compound composition. A true binary is distin-

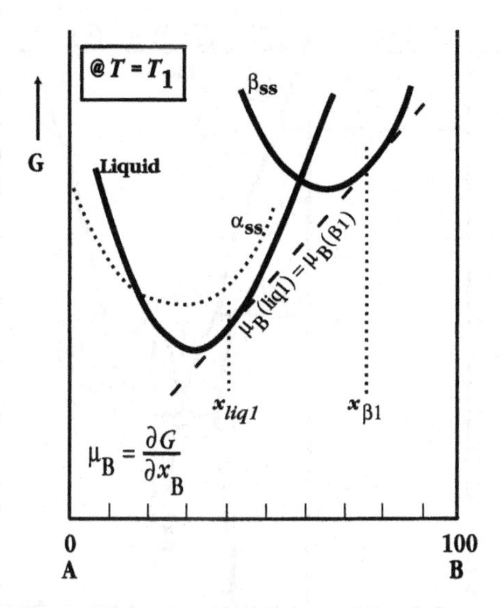

Figure 6.7 Free energy–composition schematic diagram of the system of Figure 6.6 at temperature T_1. At equilibrium, a system of overall composition between $X_{liq\,1}$ and $X_{\beta 1}$ will be composed of liquid and solid of those two compositions and will have minmum energy with respect to any other phase assembly.

guished by its two congruently melting endmembers and a single minimum liquidus temperature. Each true binary contains one or more crystallization types defined by crystallization sequence from the liquid.

Below the liquidus or crystallization temperature, various degrees of solid solution may be present. Subsolidus reaction may occur, including compound formation. Subsolidus reaction will further divide the relevant "types." The analysis of the crystallization types of the system $MgO–TiO_2$ is given in Figure 6.8.

After the first analysis of a phase diagram to determine the areas of common crystallization type, the limited number that are of practical or technological interest are readily investigated. A simple spreadsheet-type template can be prepared for each member of the "short list" if accurate phase analysis for a specific isoplethal composition is desired. Spreadsheet-type presentation of lever rule-type calculation results at various temperatures is often referred to as an "isoplethal study."

In accord with the phase rule, isoplethal studies for phase analysis are most reasonably performed with calculation at temperatures just above and just below phase transition or invariant temperatures. These calculations may also be presented in graphical format. In graphical form, the data are referred to as a "phase analysis diagram." An example is given in Figure 6.9.

For ceramics that have been crystallized from the melt or sintered with a moderate amount of liquid phase, each of the crystallization types has a distinctly

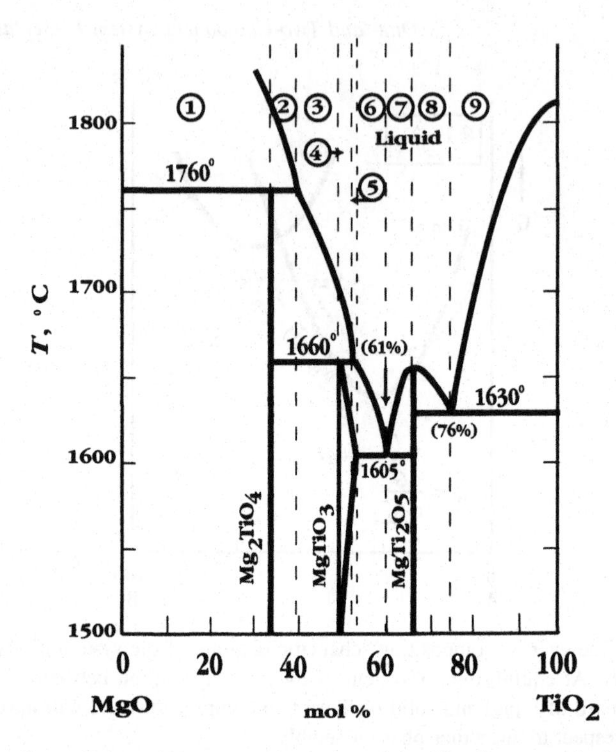

Figure 6.8 Construction of diagrams of common crystallization types for binary MgO–TiO_2. The system contains two true binary subsystems; each contains several distinct crystallization types which will exhibit similar cooling from the melt. [Phase diagram adapted from I. Shindo, *J. Crystal Growth,* **50**:839–851 (1980)].

Reaction p_1 MgO + liq(p_1) = Mg_2TiO_4
Reaction p_2 Mg_2TiO_4 + liq(p_2) = $MgTiO_3$

	1	2	3	4	5	6	7	8	9
range (mol% TiO_2)	0–33	33–39	39–50	50–53	53–5403	5403–61	61–67	67–76	76–100
liquidus	≥ 1810	1810–1760 (P1)	1760–1700	1700–1660	1660–1650	1650–1605 (e_1)	1605–1660	1660–1630 (e_2)	1630–1815
primary phase	MgO	MgO	Mg_2TiO_4	Mg_2TiO_4	$MgTiO_3$	$MgTiO_3$	$MgTi_2O_5$	$MgTi_2O_5$	TiO_2
first rxn	p_1	p_1	p_2	p_2, then ss forms 1660–1620	as in 4, ss forms 1620–1605+	e_1	e_1	e_2	e_2
2nd rxn	—	p_2	—	system remains single phase	ppt formation at $T<1605$	—	—	—	—
solidus T	1760	1660	1660	1660–1620	1620–1605	1605	1605	0630	1630
solid phases	MgO + Mg_2TiO_4	Mg_2TiO_4 + $MgTiO_3$	Mg_2TiO_4 + $MgTiO_3$	$MgTiO_3$ ss + tr $MgTi_2O_5$ ppt	$MgTiO_3$ ss + $MgTi_2O_5$ ppt (<20%)	$MgTiO_3$ ss + $MgTi_2O_5$ (eut and ppt)	$MgTi_2O_5$ + $MgTiO_3$ ss (eut and ppt)	$MgTi_2O_5$ + TiO_2	$MgTi_2O_5$ + TiO_2

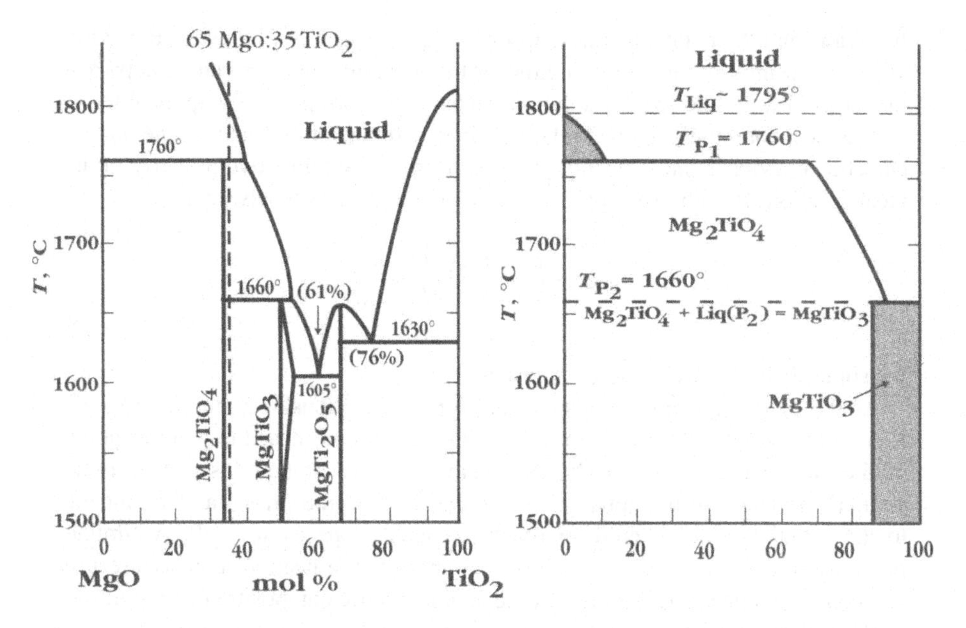

Figure 6.9 Phase analysis diagram in system 65 MgO–35 TiO_2. The horizontal axis in the phase analysis diagram (at right) is interpreted as a scale for the relative percent phase using the width of each phase field. For this analysis, expressed in mol%, the analysis must be interpreted. [Adapted from I. Shindo, *J. Crystal Growth,* **50**:839–851. (1980)].
Example: at 1600° C, the system is 88 mol% Mg_2TiO_4 + 12 mol% $MgTiO_3$

mol% phase	phase analysis	component analysis
88 Mg_2TiO_4	67 mol% MgO	0.88 × 67 = 59 mol MgO
	33 mol% TiO_2	0.88 × 33 = 29 mol TiO_2
12 $MgTiO_3$	50 mol% MgO	0.12 × 50 = 6 mol MgO
	50 mol% TiO_2	0.12 × 50 = 6 mol TiO_2

Mole proportions can then be converted to weights and wt%.

different microstructural appearance and physical performance characteristics. Within each type, a group of similar products may be obtained with some scope for property design. A seemingly minor change in chemistry that results in a major microstructural or performance alteration of a material probably "crossed the line" between neighboring crystallization types.

6.4 Heterogeneous Crystallization

Binary Univariant Interpretation

In a binary system, the regularity of curvatures seen in the univariant boundaries reflects the nature and characteristics of the solutions. Ideal and regular solution behaviors and their statistical description were discussed in Chapter 4. Behavior at low concentration of solute in either solid or liquid solutions can be approximated as ideal, while regular solution behavior is more generally applicable.

The most commonly applied relationship to estimate or evaluate simple binary eutectic behavior is the "freezing point depression," a colligative property of an ideal solution. If melt enthalpy is known for the pure materials, then

$$\ln x_B = \left(\frac{-\Delta H_{\text{melt}}}{RT}\right)\left(\frac{1}{T} - \frac{1}{T_m}\right)$$

for solute B in A (ΔH refers to A properties).

As the thermodynamic reference state for this relationship is the pure solid, it can only be used to estimate the liquidus temperature near the melting point of the pure material. Obviously, the solubility curve or liquidus can be used similarly to estimate the enthalpy of melting of the pure phase, as the tangent to the actual liquidus should approach the ideal slope as $x_B \rightarrow 0$. A similar relationship can be written to express the change in transition temperature for any condensed phase transition near the stoichiometric composition on the introduction of an impurity.

The requirement of a pure solid as reference state for application of the freezing point depression does truly limit its applicability, as some limited solubility is always noted in the solid phase. The equation yields a straight line liquidus while actual liquidus curves (for oxides) are (usually) found to be concave. Equilibrium phase compositions satisfy the condition

$$\left(\frac{\partial G}{\partial x_A}\right)_1 = \left(\frac{\partial G}{\partial x_A}\right)_2$$

or that the chemical potential of A is equal in phase 1 and phase 2. Statistical description of the solid and liquid solutions, for ideal or regular behaviors, may be stated reasonably simply, however, and can be used for either numerical or graphical solution of equilibrium between solid and liquid solutions. Experimental liquidus behaviors can be compared to calculated saturation curves for various models and used to interpret physical behaviors in terms of atomic or species interactions.

Importantly, whatever the atomistic basis for solution behaviors and regardless of the knowledge of its exact mathematical description, the invariant boundaries possess regular, constant curvatures that are associated with stable solution behaviors. This regularity allows the presentation of very limited data in the form of a meaningful phase diagram and also enables the extension of equilibrium phase boundaries for the description of metastable behavior. The experimental determination of a boundary with irregular or changing curvature versus composition is associated with highly nonideal solution behaviors, generally liquid–liquid immiscibility, or previously uncharacterized invariant reaction.

The freezing point depression estimation and a simple statistical estimation of melting enthalpy can be combined usefully to help understand the nature of the liquid phase. On melting of a solid, each atom gains three translational degrees of freedom as it is no longer constrained by lattice symmetry. Each translational degree of freedom gained contributes $1/2RT_m$ per mol of atoms in the molecular formula. Furthermore, if the liquid phase is ideal, its entropy of mixing can be estimated using a statistical mixing formula.

Recall that at equilibrium $\Delta G_{s\text{–}l}(T = T_m) = 0$, $\Delta H_m/T_m = \Delta S_m$. The entropy change on the isothermal formation of an homogeneous liquid phase can be estimated from the chemical formula and the absolute temperature of melting.

If a mole of oxide, for example ABO_3, is melted, a simple estimate of the maximum entropy of melting would be

—For lattice relaxation of 5 mol of particles (atoms)

$$\Delta S_{\text{relax}} = 5\frac{3}{2}R$$

—For ideal mixing of 5 mol of particles (atoms)

$$\Delta S_{\text{mixing,ideal}} = -5R\left[\left(\frac{1}{5}\ln\left(\frac{1}{5}\right)\right) + \left(\frac{4}{5}\ln\left(\frac{4}{5}\right)\right)\right]$$

Thus,

$$\frac{\Delta H_{\text{melting}}}{T_m} \approx 5R\left[\frac{3}{2} - \left(\left(\frac{1}{5}\right)\ln\left(\frac{1}{5}\right)\right) + \left(\left(\frac{4}{5}\right)\ln\left(\frac{4}{5}\right)\right)\right]$$

If, however, the formula were written AO_2•BO and it was hypothesized that the oxygen coordination of the cations remained relatively unchanged in the liquid, only 2 mol of particles need to be randomly mixed and the enthalpy of melting would be reduced.

—For lattice relaxation of 2 mol of particles

$$\Delta S_{\text{relax}} = 2\frac{3}{2}R$$

—For ideal mixing of 5 mol of particles

$$\Delta S_{\text{mixing,ideal}} = -2R\left[\left(\frac{1}{2}\ln\left(\frac{1}{2}\right)\right) + \left(\frac{1}{2}\ln\left(\frac{1}{2}\right)\right)\right]$$

Thus,

$$\frac{\Delta H_{\text{melting}}}{T_{\text{m}}} \approx 2R\left[\frac{3}{2} - \left(\left(\frac{1}{2}\right)\ln\left(\frac{1}{2}\right)\right) + \left(\left(\frac{1}{2}\right)\ln\left(\frac{1}{2}\right)\right)\right]$$

If no disruption of the compound molecule took place on melting, the enthalpy of melting would be simply estimated as $\Delta H_{\text{m}} \approx 3/2RT_{\text{m}}$. It is interesting to note that most ionic solids retain a large degree of bond coordination in the liquid state. The degree of molecular character retained in the liquid can often be estimated through comparison of the "ideal melting enthalpy" with the actual enthalpy, or related liquidus slope, which is observed experimentally.

In general, the liquidus becomes relatively steeper as ideal mixing is approached. Most mixed oxide liquids are fairly random mixtures of the typical coordination groups characteristic of the individual metal oxides as in the solid state. This observation lends support to the correctness of the common usage of simple oxide stoichiometric groups as endmembers in phase equilibrium diagrams that show melt behaviors.

Equilibrium Crystallization and Undercooling

The phenomena of undercooling in homogeneous phase transformation was discussed in Chapter 4, where the homologous temperature, $T_{\text{h}} = T/T_{\text{m}}$, was shown to be a useful guide to the energy available for the kinetic processes of phase nucleation and mass transport. Condensed phase reaction in a multicomponent system, particularly as described by the equilibrium univariant saturation curves, is usually observed to "lag" equilibrium when observed in real systems, for similar kinetic reasons. The physical principles underlying this behavior should be understood in the application of phase equilibrium diagrams to real systems, particularly oxide systems where atomic transport rates in the condensed phases are particularly limiting owing to stoichiometric considerations. The formalism presented here was developed by W. A. Tiller, and is an alternate but equivalent description to that already discussed for T–T–T phenomena. This analysis, although inexact mathematically, is more specific in attributing degrees of undercooling to the physical heat and mass transfer characteristics of the multicomponent system.

Figure 6.10 is a representation of undercooling parameters as determined from the equilibrium phase diagram. The driving force for crystallization from the melt is the change in free energy for the transformation which is a function of the enthalpy of transition and the equilibrium liquidus temperature of the homogeneous system, $T_{\text{L}}(C_{\infty})$, as defined by the phase diagram.

Nucleation is most generally heterogeneous, with nucleation frequency dependent on the number of available heterogeneous nucleation sites available, the

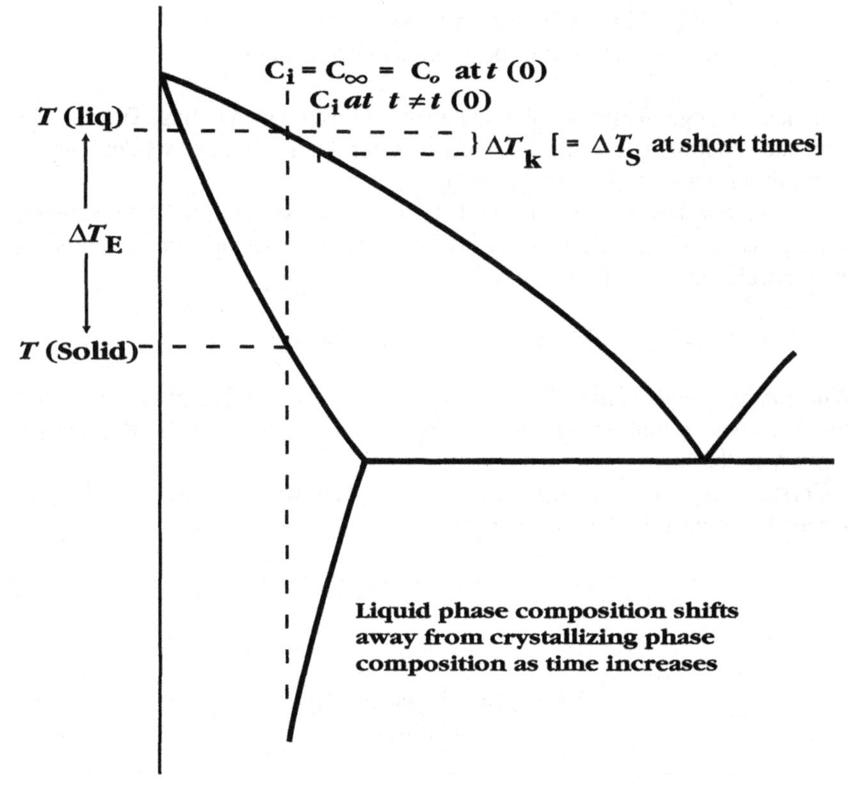

Figure 6.10 Diagram of factors used in determining crystallization parameters from the equilibrium phase diagram near the liquidus.

critical undercooling (ΔT_C), average temperature, and composition. The undercooling required is stated relative to the equilibrium liquidus or crystallization interface temperature and is the global temperature measured away from the interface at a general point in the melt. The stated undercooling is a statement of both the relative driving force as well as the kinetic lags in the system. The undercooling can be separated into several parts, as due to equilibrium, kinetic, and thermal factors.

$$\Delta T = \Delta T_S + \Delta T_E + \Delta T_k + \Delta T_H$$

where

$\Delta T_S = T_L(C_\infty) - T_L(C_i)$, the driving force for diffusion
ΔT_E = the equilibrium undercooling
ΔT_k = kinetically limited undercooling
ΔT_H = thermal transport limited undercooling

Usually a single factor will dominate the observed undercooling. The equilibrium undercooling, ΔT_E, is the difference between the liquidus and solidus temperatures of a system having composition C_i.

Random attachment of species will cause the total system energy to increase, so there is an undercooling owing to the additional driving force required to drive attachment at preferred sites.

$$\Delta T_k = T_E(C_i) - T_i = f(V, a_i, S, t)$$

Where a_i represents "lumped" attachment parameters; V is the growth velocity; and S, t relate to diffusional characteristics of the system that control species delivery to the interface.

Crystallization is an exothermic process, so there will be an effect due to the actual rate of heat dissipation in the system.

$$\Delta T_H = T_i - T_\infty = f(K, \alpha, \Delta H_f, V, S, t)$$

where

K = thermal conductivity
α = thermal diffusivity.

ΔT_H is seen to limit the growth of pure crystals, ΔT_S dominates crystal growth of oxides from the melt, while complex oxide or polymer crystallization show greater effects from kinetic factors, ΔT_k. In eutectic crystallization of metals and chalcogenides, ΔT_E is usually about half the actual undercooling observed.

In unconstrained systems where the growing crystals are widely separated in a large liquid volume, various stages of growth velocity will be observed. It is possible to calculate a value of growth velocity, through the application of statistical and dynamic methods of describing attachment kinetics, mass and thermal fluxes.

Characteristically, initially fast growth velocities slow down over time as the effects of heat transfer or attachment kinetics begin to dominate. The endpoint velocity will be determined by either the rate of mass transport or the rate of thermal transport to and from the interface. In constrained systems, the initial period driven by equilibrium forces may be vanishingly short, with apparent growth velocities limited by mass or thermal transport almost from the start.

Solute partitioning as a result of the actual growth velocity, V, is described by the ratio C_i/C_∞ which will be dependent on interface velocity, the actual partition coefficient k_i, diffusivity in the liquid, the equilibrium boundary layer thickness, the shape function, and time after the initiation of crystal growth. The partition coefficient for the solute, k_o, is defined with reference to the equilibrium phase diagram, where C refers to the solute concentration.

$$k_o = \left(\frac{C_{\text{equil,solid}}}{C_{\text{melt}}}\right)_{T_{\text{melt}}}$$

In "normal" freezing, a molten charge is frozen from one end, impurities tending to be segregated in the remaining liquid phase. Crystal growth using zone melting, the application of a moving thermal zone to a material, is the basis of many purification techniques in the formation of polycrystalline raw materials and also forms the basic technology behind containerless crystal growth methods. In several passes, significant purification can be achieved as a result of solute partitioning. Similar partitioning of impurities to the moving growth interface is characteristic of all growth from the liquid.

Non-Equilibrium Growth from the Binary Melt

Nonequilibrium growth from the melt arises when the kinetic factors for solute redistribution in the liquid and solid are insufficient to maintain equilibrium composition at the interface. For the crystallizing solid, this occurs at temperatures below $T_h \approx 0.8$ owing to mass transport limitations. Solute redistribution in the liquid may be physically constrained as a result of melt viscosity and interfacial polarization even in large volume systems. In systems with small amounts of liquid (10 to 15% or less), global equilibration may be impossible due to limited connectivity of the liquid phase as well as reduced mass transport in overlapping interfacial zones.

In nonequilibrium freezing, the true partition coefficient, k_i, depends on a number of kinetic factors in addition to the equilibrium value of k_o.

During crystal growth from a liquid, charge must also be redistributed due both from crystal bond formation and local nonstoichiometry in the liquid phase. A boundary layer develops, as shown in Figure 6.11. The liquid must be characterized as an electrolyte, particularly its dielectric properties, to fully understand this effect. A macropotential $\Delta\phi$ results from the characteristics of the electrolyte and the concentration of charged species at the interface. The potential represents a driving force for recombination of ions to form neutral species. The partition function of solute species between liquid and solids is a strong function of the developing macropotential.

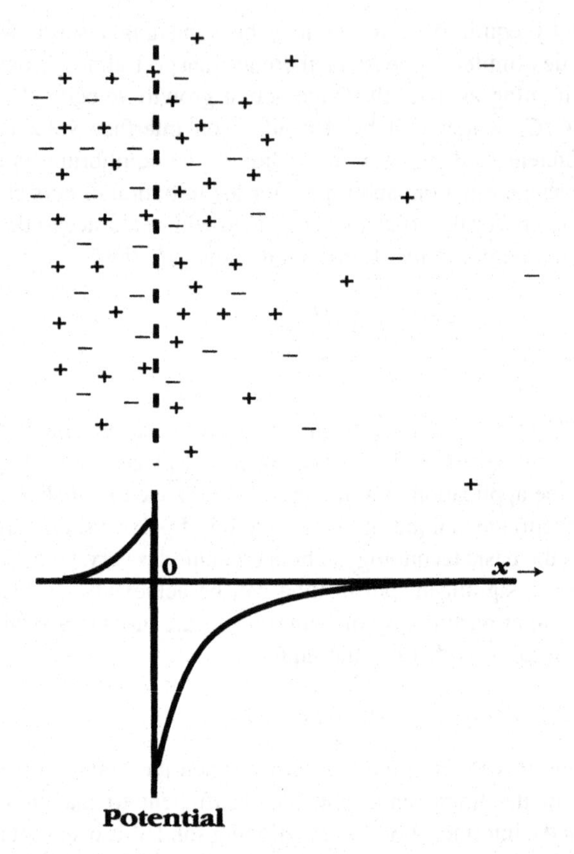

Figure 6.11 Schematic boundary layer development due to charge distribution at a moving growth interface. Liquid or solid solution possesses net zero charge away from the interface.

Significant polarization at the interface can effectively stop growth in constrained systems where solute redistribution is not possible through convection. Static boundary layer segregation at the growth interface restricts the free mobility required for equilibration. As a local field effect, however, this is not incorporated into the phase rule.

The partition coefficient cannot be simply derived for the case of liquid phase sintering. During sintering, it is likely that the liquid phase that develops is physically thinner at contact points than the classically derived diffusion boundary layer thickness for crystal growth from an unconstrained melt. High degrees of impurity segregation at grain boundary areas are found in the latter stages of sintering. Solute partitioning, even due to impurity segregation during solid state grain growth, can result in the formation of a chemically complex boundary composition phase that may melt (possibly causing slump of overfired ceramics).

Segregation Phenomena

Microstructural phenomena that commonly result from segregation during quasi-equilibrium (or unconstrained) and nonequilibrium constrained crystallization growth from the melt include "coring" (or "zoning") and grain boundary segregation leading to glassy or crystalline residual grain boundary phases after cooling. On annealing, nonequilibrium solid phases may exsolve excess solute, forming internal or grain boundary precipitates. "Annealing" processes may be intentional heat treatments or may result from device applications at $T_h \geq 0.5$.

Zoning and coring are the result of crystallization in which the interfacial crystallizing composition is close to equilibrium, but where slow diffusion rates in the crystalline phase do not allow for internal compositional equilibration. This situation implies that heat and mass transport in the liquid solution phase are not limiting factors, normally true in the case of metals.

Zoning is also observed in the formation of precipitates from a saturated solid solution, although equilibrium is seldom maintained in either solution or precipitate phase except in the case of very high transport rates. Void formation or solute depletion zones are commonly observed in cases of solid–solid precipitation, leaving a persistent "trace" for investigation, similar to that used in classic diffusional studies.

Zoning is observed in oxides, particularly under geological conditions for silicates, where changes in the "original" compositional bands can be used to characterize past volcanic and other upheavals.[5] In oxide refractory systems, particularly fused-cast refractories, individual crystallites may exhibit zoning "as cast." When refractories are used for extended times under static thermal gradient conditions, alteration or zoning of individual grains and possible precipitate formation is a similar indicator of local thermal conditions.

Interfacial segregation during freezing occurs in constrained systems and where the liquid phase is viscous, limiting mass transport. During forced cooling, component segregation may lead to local supersaturations which favor the appearance of a nonequilibrium phase. A simple hypothetical case is presented in Figure 6.12. If cooling is sufficiently rapid, the nonequilibrium phase could be maintained in the microstructure and affect further crystal or grain growth.

Interfacial segregation is also observed as grain boundary segregation, where an impurity not necessarily included in the system description is excluded by the growing crystal and becomes concentrated in the remaining liquid. Segregation coefficients of 100 to 10,000 are not unreasonable, resulting in rapidly increasing impurity concentrations in the residual grain boundary melt. Depending on melt viscosity and complexity, the residual liquid may form a glass or produce one or more crystalline phases on further cooling. Any precipitate would be character-

---

[5]See *Kinetics and Equilibrium in Mineral Reactions*, edited by by Surendra K. Saxena (1983), Springer-Verlag, NY.

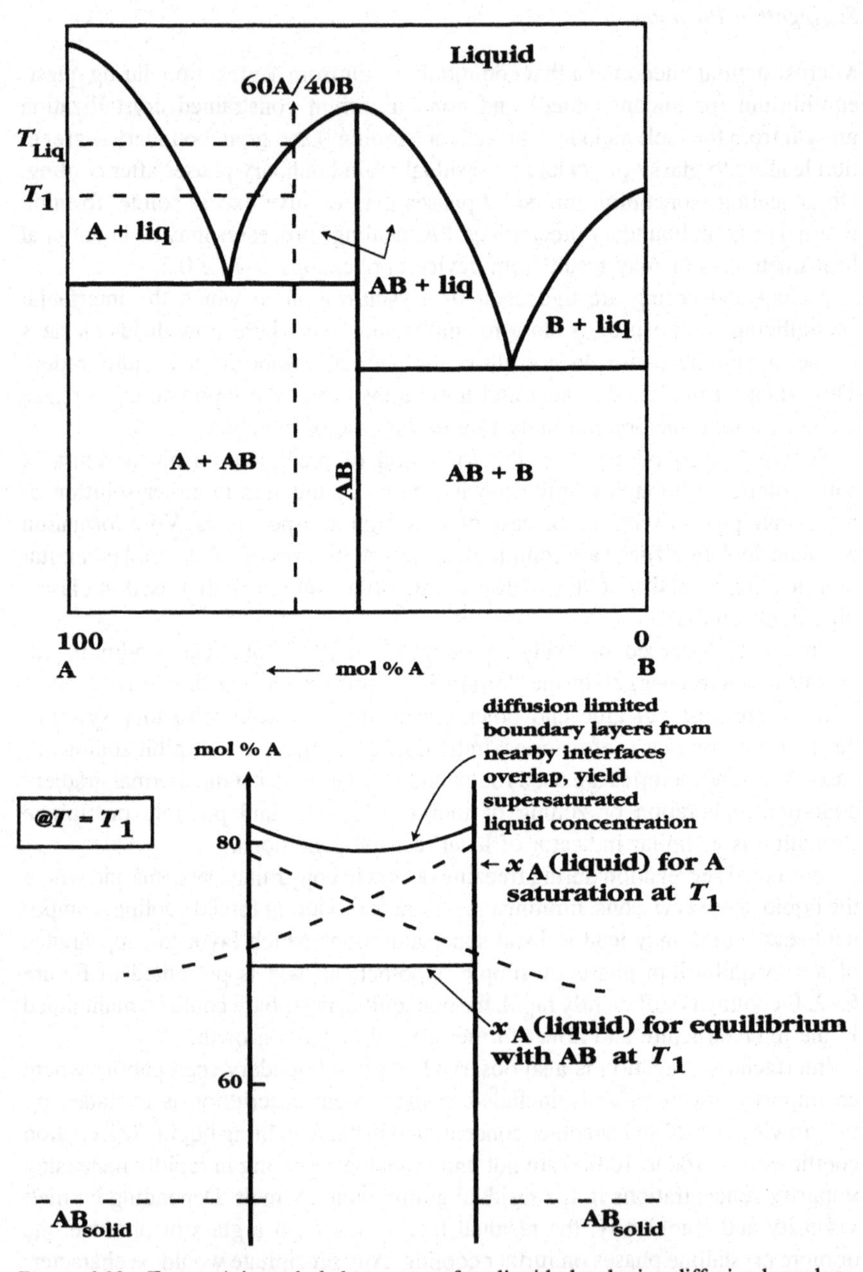

Figure 6.12 Excess A is excluded to near surface liquid, developing diffusion boundary layer near AB precipitate. In constrained systems, boundary layer thickness exceeds particle separation, resulting in metastable A-supersaturated liquid.

istic of the local interface concentrations of this more complex n-component system, and would be expected to have at least a metastable compatibility relationship with the primary crystal phase. (Sintering aids are often effective owing to their influence on the nature of the final grain boundary phase and its properties. See Chapter 8.)

Grain boundary precipitation can also occur during annealing of a homogeneous but supersaturated solid solution. In the subsolidus region, favorable entropy considerations are overwhelmed by the enthalpy of solid solution formation with the net result of more limited equilibrium solute concentrations. Annealing of a supersaturated solid solution will result in a microstructure that is consistent with the characteristic atomic mobilities and the structural characteristics of potential stable and metastable precipitate phases.

In classic alloy precipitate hardening, an alloy (composition S) is cooled rapidly. A small amount of second phase B forms on cooling, but the system is not allowed to equilibrate. Obviously, from the earlier discussion on T_h, the two solids will coexist indefinitely at a low temperature, $T < 0.5T_h$, owing to kinetic limits on equilibration.

The phase diagram is given in Figure 6.13. The system if heated to $T \geq T_1$ will revert to a single-phase solid solution α; heated to any lower temperature, the phase compositions will gradually shift toward equilibrium values with exsolution of component B from the α_{ss}. The spatial scale of precipitation will reflect the time and temperature through the diffusion coefficient of the solute relative to the grain size and also the availability of nucleation sites. In large-grained

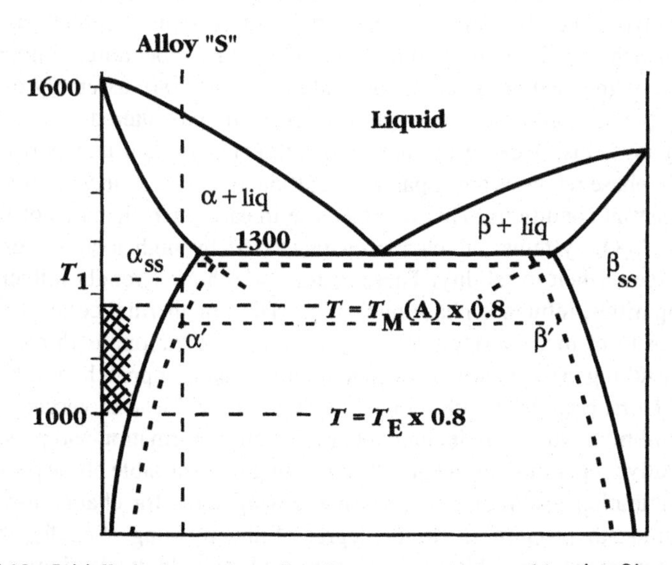

Figure 6.13 Initially homogeneous alloy S separates to metastable $\alpha'_{ss} + \beta'_{ss}$ on cooling. When heated into the annealing temperature range, excess B will precipitate.

metals, precipitates are seen to decorate dislocations after short-term annealing, whereas in fine-grained oxides, precipitation at pore surfaces or at the grain boundaries is more likely.

In systems where the equilibrium precipitate phase has a complex structure or stoichiometry, supersaturation of a solid solution may persist metastably or a metastable defect solution phase may develop. Metastable precipitate formation is also likely in multicomponent systems where transport rates are very unequal.

6.5 Binary Metastability

Metastable coexistence of phases is routinely observed in materials processing, generally owing to the effects of sluggish reaction, phase nucleation, or growth kinetics. True metastability satisfies the Gibbs equation, that the chemical potentials of the components are equal in the coexisting phases. However, these phases are not the lowest energy phases that could satisfy that requirement, leaving the system unstable with respect to a fluctuation toward the true equilibrium state. The equilibrium phase diagram shows the lowest energy assemblage of phases that satisfy the Gibbs requirement.

True metastability is a relatively high temperature phenomenon and will be seen to satisfy the phase rule. A metastable phase diagram is prepared using the equilibrium phase diagram and is characterized by the complete absence of an equilibrium phase. The diagram is used to investigate the "what if" questions about the behavior of a current phase assemblage. The current phases may be representative of equilibrium, as a melt that will be rapidly cooled; may have been at or near equilibrium at some prior time and temperature, similar to the case of the segregated alloy above; or could be an arbitrary assembly of phases typical of powder processing that are to be reacted and sintered.

In some systems, because of characteristic nucleation and transport kinetics, metastable phase relations are apparently reproducible over a range of experimental or industrial conditions. This was the case in early investigations of the SiO_2 and SiO_2–Al_2O_3 systems, which are characterized by high melt viscosity and some degree of melt instability. These factors, which are strongly influenced by minor impurities, influence phase nucleation. The low thermal conductivity and high viscosity of these systems also influence crystallization, as discussed previously, resulting in an "equilibrium undercooling" in the approach to equilibrium. Kinetic factors, combined with a finite cooling rate, will lead to varying degrees of disequilibrium which can include metastable phase formation and persistence.

The relative separation from equilibrium owing to metastable phase persistence or supersaturation results in an increasing driving force for change toward the equilibrium state, proportional to the degree of undercooling. (See discussion of T–T–T phenomena.) Possible supersaturation is limited by nucleation and transport rates, as a fluctuation toward equilibrium has an increasing driving force for persistence and growth. As undercooling increases, the likelihood of metastable phase nucleation also increases. Typically, a metastable phase that is experimentally observed requires less physical alteration of the current state to achieve—a displacive transformation or the ordering of a relatively mobile species—requiring no significant nucleation step. The driving force for metastable phase transformation is less than that toward equilibrium but the state is more readily achieved. Any metastable phase formation will reduce the effective driving force toward true equilibration of the system.

Metastable Crystallization from the Melt

Metastable phase diagrams are constructed using the equilibrium univariant curves. These curves are (or can be) fully described through the thermodynamic functions which are regular and without singularity for each phase. The univariant curve can be extended, therefore, through the equilibrium triple point (or invariance) which, under equilibrium conditions, would have marked the appearance of a new phase. Figure 6.14 shows examples of metastable diagrams that may be constructed for problem solving in nonequilibrium crystallization processes.

In the first example, the equilibrium phase C_2D does not nucleate. What will be the composition of the first metastable solid phase? The metastable diagram is generated first by extending the equilibrium liquidus, C_{ss} + liq, through the equilibrium triple point. The B_{ss} + liq univariant can be similarly extended; the two intersect at a metastable eutectic, e', defining the metastable solidus. The upper saturation curves for each solid solution are extended to meet the solidus at the temperature of metastable invariance. The lower saturation curves are then displaced laterally, parallel to the original curves, to the new metastable invariant solid solution compositions. In this example, it is seen that the solid solution C′ would be expected to crystallize metastably from the supercooled melt X_1 at temperature T_1, while the solid solution D′ would be predicted from supercooled melt X_2 at T_2.

In practice, either of these metastable products could serve as heterogeneous nuclei for the equilibrium phase if the metastable event occurred at a temperature for good mobility in the melt. If the nuclei affected the equilibrium growth habit, it is possible that the metastable eutectic composition e' would mark an alteration in microstructure for rapidly cooled melt compositions between P and E. If the nucleation event simply preceded equilibrium phase growth, the final microstructure would not present significantly different phase proportions than for the equilibrium prediction, but the relationship between phases or their morphology may reflect their nonequilibrium excursion.

The equilibrium peritectic reaction at $0.67 > X_D > P$ is a dissolution–reprecipitation phenomenon. At a finite cooling rate, reaction will be limited by mobility of either component in either the solid or liquid phases. The most likely outcome is an incomplete reaction of the primary C_{ss} crystals, with the

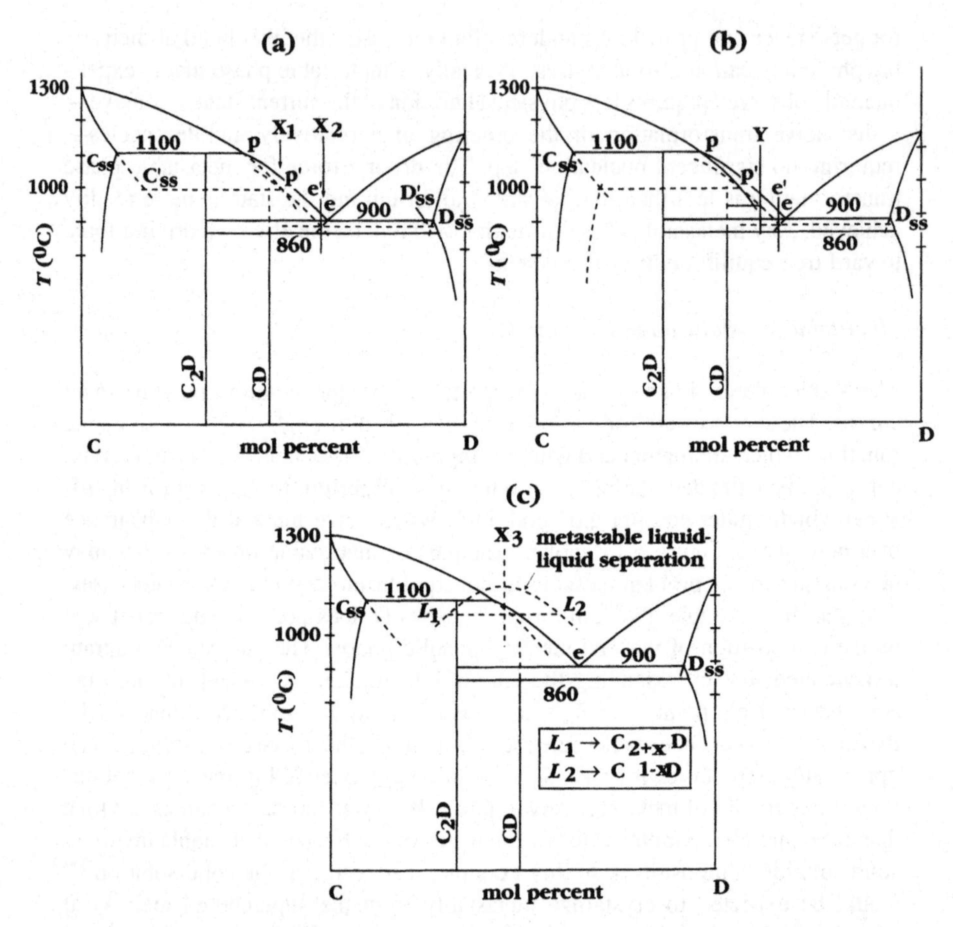

Figure 6.14 (a) Metastable extension of liquidus curves generates metastable eutectic at *e′*. Metastable crystallization of liquid X_1 yields C_{ss}' while metastable crystallization of X_2 yields D_{ss}'. (b) Metastable crystallization of phase CD from liquid Y indicates metastable peritectic reaction P′. Metastable solid–solid equilibrium C_{ss} + CD would also be possible in this construction. (c) Metastable liquid–liquid phase separation of supercooled liquid X_3 could lead to crystallization of metastable solid solution compositions.

stoichiometry of a metastable crystallizing phase from the C-saturated liquid being characteristic of the surface boundary layer composition, a metastable solid solution phase. If we assume that the compound stoichiometry does represent a low-energy configuration for these components, the metastable phases formed through any scenario will have a structure similar to the stable equilibrium phase or to another solid phase known to form in the system. In this hypothetical case, metastable stoichiometries from C-excess "C_2D" to C-deficient "CD" could be observed in practice. A "possible" phase stoichiometry can be determined only through experimental evidence if it does not have an equilibrium phase field. Phase nucleation is a statistical event, so many such phases may nucleate within an experimental mixture, but the equilibrium phase or phases will eventually nucleate and grow at the expense of transient metastable phases if mass transport kinetics allow.

In a low-mobility (but not glass-forming) liquid, the metastable crystalling phase may more closely resemble a simple molecular group characteristic of the liquid or the liquid's composition than an equilibrium stoichiometry. The relatively shallow liquidus slope that is characteristic of covalent oxide melts is indicative of nonideal liquid behavior, often tending toward separation or "weak" compound formation, or liquid–liquid immiscibility. What is seen as a stable compound on the phase diagram may be a result of this instability. The availability of a preexisting or heterogeneous nucleus could result in metastable precipitation from the nonideal liquid, similar to the case of the glass ceramic. Very careful experimentation is required to determine whether the compound formation (peritectic reaction) or liquid–liquid phase separation (monotectic formation) is truly equilibrium.

In viscous glass-forming liquids, there may be a detectable change in temperature dependencies of thermal conductivity and characteristic molar volume near the equilibrium liquidus, which reflect local structure reorganization toward more stable configurations. In silicate glasses and glass ceramics, these metastable structures are reflected in characteristic devitrification phases, crystalline phases that can be formed only from a glass precursor, such as "devitrite," β-spodumene, and β-eucryptite.

Metastable crystallization is difficult to describe other than as inferred from microstructural and microprobe-type analysis. Liquidus determination on cooling from high temperatures is usually indirect, through detection of the heat evolved on crystallization via a probe immersed in the melt or embedded in the containment. The thermal conductivity characteristics of system surroundings and probe limit detection of small amounts of crystalline phase formation, thus small degrees of undercooling and metastable phase crystallization are easily obscured by experimental uncertainties.

Stranded Phases and Instability on Heating

The simple diagram in Figure 6.14(a) can also be used to demonstrate the behavior of a mixture of solids C and D on heating. The metastable phase diagram generated above can be used to show that the mixture of C and D can exist in metastable equilibrium in the absence of phases C_2D or CD, with a tendency for increased solid solution formation, up to the temperature of the metastable eutectic. Above this temperature, no compositions of the two solids are compatible and metastable melting is predicted.

Often, the metastable eutectic temperature marks the rapid onset of stable phase formation, as an enormous driving force exists toward the formation of a compound from the "liquid-like" state. Preexisting nuclei, due to structural similarity between either of the supersaturated solid solutions, from prior calcination heat treatments or mechanical damage could also have led to earlier phase development. For practical processing with "stranded" initial phases, controlled nucleation through calcination for solid-state reaction is essential for reproducible transformation to the stable phase, especially where more than one phase stoichiometry is possible.

It should be remembered that the phase diagram indicates only the stable low-energy phases of a system. Ordered solid solutions and alternate phase stoichiometry characterize metastable equilibria in systems of great practical use, such as the Fe-C and zirconia electrolyte and glass-ceramic forming systems. True equilibrium phase development may require an extended period of observation, sometimes a million or more years for solid-state transformations. Thus, a subsolidus phase diagram may be experimentally reproducible and still not represent true equilibrium. Secondary criteria other than physical persistence must be applied to establish equilibrium compatibility.

Pressure and Solute Effects

Condensed phase equilibrium diagrams are presented for a fixed value of the mechanical pressure, most usually atmospheric pressure for an unconstrained system. In practice, the effective mechanical pressure or stress state may vary widely in a ceramic during processing and application. Applied pressure or mechanical stress can effect solubility and phase transition behaviors. In the absence of an experimental equation of state, the effect of pressure on the equilibrium phase diagram is best discussed using LeChattelier's rule.

LeChattelier's rule is a corollary relationship to the Clapeyron equation and expresses in qualitative terms the effects of partial molar quantities associated with compositional, thermal, and mechanical stresses. Simply stated, a system will respond to an incremental external stress in such a manner as to minimize the change in free energy experienced (similar to the inertia of a mechanical system which, when at rest, tends to stay at rest). Whereas the Clapeyron equation is applicable only to congruent, homogeneous transition behaviors, LeChattelier's rule is a guide to the behavioral tendencies of complex systems in which the compositions may be changing or for which exact knowledge of thermodynamic properties is impossible. An observation that is "in accord with LeChattelier's rule" will also be consistent with the requirements of equilibrium thermodynamics. LeChattelier's rule complements the phase rule and supplements our limited ability to directly quantify the behavior of chemically complex systems.

The simplest examples relate to the concept of P–V work. For example, LeChatellier's rule implies that a denser one-component phase will be more favorable at increased pressure or would imply that the solubility of a large atomic size solute would decrease under those conditions (both favored outcomes resulting in smaller molar volume). In accord with this "rule," an increased average coordination number in the stable high-pressure phase is noted in many systems. There are situations in which the electronic configuration of the cations will be shifted in the high-pressure phase for a net reduction in molar volume as well; for instance Fe^{2+} in the low spin configuration is smaller than in the high spin configuration and thus would be more favored as pressure increases.

A constrained system subjected to an increased temperature will be subject to mechanical and thermal stresses which may be minimized through structural transition. Congruent polymorphic transition temperatures are calculated where possible using the Clapeyron equation, whereas solution effects may be deduced using LeChattelier's rule. If, for example, increased pressure is known to decrease the phase transition temperature of the pure phase, the incorporation of a larger solute in a constrained phase should also lead to a reduced transition temperature.

6.6 Rules for Diagram Construction

Phase diagrams are most usually constructed using limited experimental data for phase occurrence, composition range, and stability. For the kinetic reasons explored briefly in the preceding sections, experimental data will generally represent some degree of nonequilibrium or metastability. The phase diagram constructed from these data cannot be defined exclusively by experimental results, but also using the phase rule, a logical physical description of each transformation and solution effects, and rules for the construction of the saturation curves and their metastable extensions. An experimental phase diagram postulates equilibrium and guides further experimentation. This discussion on diagram construction, specific for binary oxides, can readily be extended to higher order systems.

Invariant Behaviors

A nominally binary system of oxides is composed of at least three chemical elements—two metals and oxygen. The maximum number of components is three if oxide stoichiometry is not effectively fixed, but generally condensed phase equilibrium is described with $C = 1$ or 2. Congruent phase transformation, such as melting of a compound or solid solution and azeotropic vaporization, is associated with a temperature arrest. A temperature arrest, or invariance, associated with the coexistence of two phases at a fixed pressure is the hallmark of a one-component system. Therefore, it is seen that solid solution limits go to zero at the congruent melting point for endmembers and for congruently melting or critical solution compositions.

The behavior of solid solutions at a transition must reflect the logical behavior of the free energy–composition function (chemical potential) of each phase. As composition varies, the partial molar volume varies distinctly for the high- and low-temperature structures; therefore the temperature at which the two polymorphs are compatible must vary with composition. It is possible that a congruent composition will exist; away from this composition, the transition must be described as a univariant saturation with a finite slope and curvature in T–X space. The curvatures of the saturation curves of both polymorphs must go to zero at the composition of the congruent transformation.

Univariant Behaviors and Metastable Extensions

Compatible phase saturation curves must also reflect the inflection of chemical potential function that occurs at the polymorphic transformation. An inflection will mirror the phase transition behavior. As $C = 2$ (away from the singular congruent composition), the coexistence of three phases is an invariance. Only an unique composition of the high-temperature polymorph and a similarly unique composition of the low-temperature polymorph can coexist with a third compatible phase.

The behavior of saturation or invariant curves at a triple point is governed by the rule of metastable extensions. In a one-component system undergoing a change of state, the rule can be logically demonstrated using the Clapeyron equation to compare the slope of each curve as it enters a triple point. It is readily shown that the extension of each curve must lie between the other two.

For multicomponent diagrams in T–X space, a similar behavior holds but cannot readily be proven as a physical law. The author has observed only one case in which this rule was violated, in a case where a structural phase transition in one of the phases occurred very close to, but not at, the conditions of the diagram. In that case, the diagram conditions would be deemed inappropriate, as the assumption for simple two-dimensional presentation of multicomponent equilibrium is that the energetic functions of the phases are continuous and without singularity in the pressure–temperature region of the section presented.

Barring such unexpected singularities, the rule for metastable extensions does apply to all triple points, for univariant curves or their projections in the isobaric plane of the diagram. The univariant saturation curves of the T–X plane are easily recognized as isobaric cuts in a divariant surface in P–T–X space. Similarly, the phase compositions at their intersection, such as the eutectic liquid composition or critical phase composition (such as at peritectic decomposition or for solid solution transformation discussed above) belong to a true univariant curve in P–T–X ($F = C - P + 2$) space which can be projected into the plane of the diagram. This projection would be expected as almost vertical if the free energy–composition curves of the three phases do not have strong pressure dependencies, but could have finite slope compatibilities involving nonideal solutions or phases where the compressibility is a strong function of composition. Metastable extensions which would imply an extreme slope are not often encountered. Implied nonideality should be reflected at other transition point behaviors or confirmed through other property measurements to be accepted.

The variation of univariant saturation curves must also be physically logical above and below a transition. Even in cases where the metastable extensions appear correct, the indicated curvatures can be illogical. For instance, solid solution saturations generally decrease on cooling, owing to the decreasing influence of entropy versus enthalpy for solution formation. Similarly, a solid does not melt on cooling or a liquid freeze on heating. A slope that implied otherwise would not be logical, and is either a simple mistake or could possibly indicate an undocumented transformation or phase formation.

The proof of any diagram presentation is the clear and unambiguous description of isoplethal behaviors through any and all reaction and transformation processes.

7

Three or More Component Equilibria

7.1 The Ternary System

The maximum variance of a condensed, isobaric three-component system is 3. In a two-dimensional representation, the variance that can be described is restricted to values of 0, 1, and 2. The usual presentation is of a projection of the saturation surface of the single-phase liquid. An area ($\Phi = 2$) is a two-phase region; a line or curve ($\Phi = 1$) is a three-phase equilibrium, and the intersection of lines at a point ($\Phi = 0$) represents the isobaric invariance of four-phase equilibria. A projection of the liquidus onto a single plane is unambiguous if the solid crystallizing phase compositions are reasonably constant, as unique isothermal sections can be determined from the diagram information.

Typically, subsolidus diagrams are schematic and nonisothermal. The typical subsolidus diagram is used to calculate the nominal equilibrium phase proportions expected for a system of known composition, and is very useful (see discussion of Figure 7.3), but cannot be used to understand reaction processes or metastability. Subsolidus ternary *equilibrium* diagrams can be prepared for isothermal conditions, and are essential in systems with significant solid solution formation.

A ternary liquidus diagram is a projection of the primary liquidus saturation isotherms, defining the primary phase fields. The intersections of these primary fields define the univariant curves or boundaries, in which two solids coexist in equilibrium with a liquid. The univariant or boundary curves intersect at invariant points of four-phase equilibrium, either eutectic or peritectic reaction. Subsolidus reaction equilibria are not explicitly presented, but are implied as the persistence of the three solid products of the invariant reaction.

The "system" has a known composition, and is observed isothermally. A standard presentation of composition is using triangular axes (see below) and simple oxide endmembers. Temperature is a vertical axis out of the plane of the paper and it is assumed that only the composition of the liquid phase varies with temperature while the compositions of the solids may be approximated as constant. If this is a reasonable assumption, the interpretation of the diagram between the highest melting point and lowest eutectic temperatures is unambiguous using straightforward graphical elements termed "Alkemade lines."

The Alkemade lines of a ternary system are special joins constructed between crystalline endmembers that can coexist with a liquid phase over a range of temperatures (univariant equilibrium). These two phases coexist with the liquid at the temperatures along a boundary curve separating their primary crystallization fields and will be associated in a ternary triple point reaction. Importantly, Alkemade lines will never cross and divide the diagram into discrete regions. The construction of the various Alkemade lines in a typical ternary is demonstrated in Figure 7.1.

A *true binary* is an Alkemade line connecting the composition of two congruently melting phases that crosses only the liquidus boundary between those two phases. A ternary composition that is located on an Alkemade line that is a true binary is described with one less compositional degree of freedom. These binary compositions will crystallize completely in a binary eutectic reaction. The eutectic liquid composition is the intersection of the Alkemade line with the ternary univarient boundary.

A major binary subsystem of a ternary is usually a colinear collection of such Alkemade lines. The endmembers must be appropriate components to describe all the phases that may occur along the length of the true binary.

A *pseudobinary* is an Alkemade line connecting two phases, at least one of which is usually incongruently melting, which crosses more or less than the single solid–solid–liquid boundary of its own two solids. A pseudobinary diagram can be generated for this line which is a representation of the vertical section, or crystallization paths, of that series of compositions in the ternary system. For all compositions belonging to the pseudobinary, the Alkemade line again defines the final crystallized phase proportions after ternary peritectic reaction. A pseudobinary diagram does not necessarily obey the phase rule for a binary system and cannot be used to calculate phase proportions or to uniquely determine compatible liquid phase compositions.

(The term "pseudobinary" is also used to describe any arbitrary line of compositions in a multicomponent system. But an arbitrary line is not an Alkemade line. The pseudobinary diagram for an arbitrary line can be valuable for the presentation of reaction and compatibility information for a series of related compositions as might be seen in manufacturing.)

Three Alkemade lines will define an *Alkemade triangle*. An Alkemade triangle defines the four-phase invariance of a ternary, as all three crystalline phases coexist with a liquid at a composition point and critical temperature. If that invarience point is within the Alkemade triangle, the four-phase invariance is a *eutectic* and the compatibility triangle is in itself a true ternary system. If the

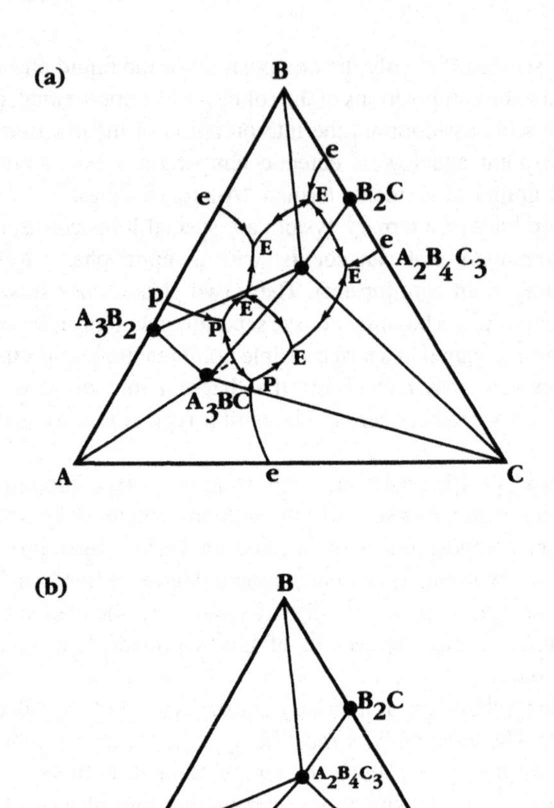

Figure 7.1 (a) Liquidus diagram of hypothetical system A–B–C with two ternary phases. A_3BC melts incongruently; $A_2B_4C_3$ melts congruently. True binaries within the ternary are B–$A_2B_4C_3$, B_2C–$A_2B_4C_3$, and C–$A_2B_4C_3$. Pseudobinaries are B–A_3BC, A–A_3BC, A_3B_2–A_3BC, and C–A_3BC. The system includes two ternary subsystems: B–$A_2B_4C_3$–B_2C and C–$A_2B_4C_3$–B_2C. (Liquidus isotherms are not shown, arrows show direction of falling liquidus temperature in accord with the Alkemade theorem.) (b) Subsolidus system of (a), showing the seven compatibility triangles defined by the Alkemade lines.

point of invarience is at a composition outside the triangle, one or more *peritectic* reactions will occur during equilibrium crystallization from the melt.

For a system having simple crystallization (not involving extensive solid solution), determination of the Alkemade lines is the first and most necessary step in understanding crystallization from the liquid. A single system composition is found to lie either within an Alkemade triangle or on an Alkemade line, and usually it will also be in a divarient region.[1] A composition on the Alkemade line is a member of the binary or pseudobinary section defined by that line. A position in the divariant region indicates an intersection or "piercing" of the liquidus at the temperature of saturation with the primary crystalline phase, defining the onset of crystallization during cooling. The location of a system's composition within an Alkemade triangle identifies the final liquidus reaction and reaction temperature. The exact position in the triangle gives the equilibrium "balance" point of the three solid phases in the subsolidus region (providing no subsolidus decomposition reactions occur). In the temperature range between these extremes of first and final crystallization, the phase equilibrium is determined at each temperature using the isothermal tie line or tie triangle which includes the composition point.

The liquidus isotherms are the projections of the single-phase saturated liquid compositions at those temperatures. In a divariant area, each isotherm represents the endpoints of the tie lines: "liquid + primary crystal." The tie lines are generally constructed schematically and are seen to radiate evenly from the composition of the crystallizing primary phase which is approximated as a point, even in cases where limited binary solid solution (along the Alkemade lines) is indicated in the diagram. The lever rule is applied in the same manner as in a binary or quasibinary system for simple mass balance between two compatible phases.

The isothermal saturation curves of the primary crystallization regions intersect at the univariant curves, forming an isothermal compatibility triangle. The triangle is a plane in the volume "liquid + primary crystal + secondary crystal," where one side of the triangle is the Alkemade line corresponding to the two crystallizing phases and the other two sides are constructed from the liquid composition point where the liquidus isotherm intersects with the univariant boundary.

Univariant curves are of two types, cocrystallization and reaction boundaries. In the first case, the reaction "liquid = primary crystal + secondary crystal" proceeds at all points on cooling. In the second case, the reaction "liquid + primary crystal = secondary crystal" proceeds as the system cools. To determine the type of reaction occurring isothermally, it is necessary to estimate the tangent to the univariant curve at the isothermal temperature of interest.

The tangent to the univariant curve has interesting properties which are particularly valuable in crystallization studies. Simultaneous crystallization or reaction requires that a continuous balance of mass and energy be maintained. As a result, as temperature varies during cocrystallization, the tangent to the univariant curve will intersect a point on the Alkemade line. Application of the lever rule at the point of intersection yields the proportions of phases that have crystallized between T and $T - \delta T$. In the case of a reaction at the univariant boundary, the

[1]The composition might lie on a univariant curve or an invariant point, but that is not a general case.

tangent will not intersect the appropriate tie line of the ruled surface at all (or possibly any) of the temperatures presented.

Cocrystallization typically proceeds to the eutectic or peritectic reaction. A "reaction boundary" is seen to be an extended peritectic reaction that proceeds only until the reactant is consumed. Once a reactant is exhausted, the system is no longer univariant and a new divariant equilibrium must be defined. Preparation of a crystallization map for the ternary system is useful to pinpoint these regions and their characteristic reactions (see below).

The Alkemade Theorem

The intersection of the Alkemade line (or its extension) with its specific boundary curve or that curve's graphical extension[2] indicates a relative maximum in temperature along that boundary. This is the "Alkemade theorem." The Alkemade theorem is useful for the interpretation of complex liquidus reaction sequences, particularly involving incongruently melting ternary compounds. The Alkemade theorem is also used in the logical construction of the liquidus from limited experimental data.

As the Alkemade lines define subsolidus equilibria, they can be determined from careful solid-state experiments. Consistent finding of three compatible phases in a composition region suggests an Alkemade triangle. Melting experiments can then be undertaken to determine the lowest melting compositions, which should correspond to peritectic or eutectic reactions. Using the Alkemade theorem, a tentative liquidus diagram can be constructed. For a eutectic, no liquid remains after reaction so all three univariant slopes are downward (or away from the Alkemade lines) toward the triple-point liquid composition. For a peritectic reaction, one boundary must slope away or outward from the triple point. Quenching experiments are used to determine the crystallizing phases in complex systems.

The Alkemade lines are also used to check the presentation of the system's crystallization reactions. Each boundary on a ternary diagram is generally labelled with an arrow that indicates the trend or slope of remaining liquid compositions as temperature decreases and univariant reaction proceeds. The Alkemade theorem is a simple check on the consistency of a published diagram against logical and typographical errors.

Extended Solid Solutions

Crystallization of solid solutions must be presented in discrete isothermal sections for meaningful interpretation of phase equilibrium between the temperatures of

---

[2]A simple imaginary line extension to the Alkemade line is made, corresponding to the true metastable extension of the boundary, but without as much care being needed with regard to curvature. Similarly, the Alkemade line may be extended, if required, to intersect its boundary for application of the Alkemade theorem.

initial and final crystallization. Within each isothermal section of the liquidus, unique tie lines define the compatible liquid solution and solid solution compositions for divariant and univariant equilibrium. Subsolidus regions may be characterized by large quasibinary regions, as well as three-phase compatibilities.

In most experimentally determined phase diagrams, the tie lines joining compatible solution compositions are "schematic." If both of the compatible phases are ideal or regular solutions, the tie lines can be assumed to lie in a regular array between the two saturation curves: from centerpoint to centerpoint, from quarterpoint to quarterpoint, etc. An experimentally observed deviation from such regularity would indicate that the isothermal section lies near the critical temperature for single-phase stability of one or both of the solid solutions, or would indicate highly nonideal behavior of one or both solutions.

As in simple crystallization, each isothermal section has characteristic regions of divariant and univariant equilibrium. The location of the single system composition within a region will determine the appropriate tie line or tie triangle to be used to determine the mass balance between compatible phases.

Metastable Extensions

The univariant curves in a ternary diagram can be extended through a triple point, as discussed in binary systems, and the rules for metastable extensions apply. For the simple liquidus diagram, metastable extension can be used to estimate reaction temperature, or maximum metastable compatibility, between starting phases as demonstrated for the binary. The limited two-dimensional diagram format is difficult to extend out of the plane of the paper. A reasonable estimate for the maximum temprature of metastable coexistence of incompatible solids would be the lowest stable eutectic temperature located within the tie triangle of the original phase compositions.

Metastable phase diagrams as would result from the kinetic limitation on nucleation of complex ternary phases can more readily be generated, through extension of the univariant curves through the invariant points that define the primary field of the kinetically unfavorable phase.

In subsolidus isothermal diagrams, the bounding curves of saturated solid solutions are also univariant boundaries. It is noted that the path of the metastable extensions of the single-phase saturation surfaces relate to the relative stability of the solution, as indicated in Figure 7.2.

7.2 Ternary Phase Compatibility and Mass Balance

The interpretation of ternary phase equilibrium for a single-phase composition involves accurate graphical manipulation in triangular compositional coordinates. In each successive isothermal section, tie lines and tie triangles define phase

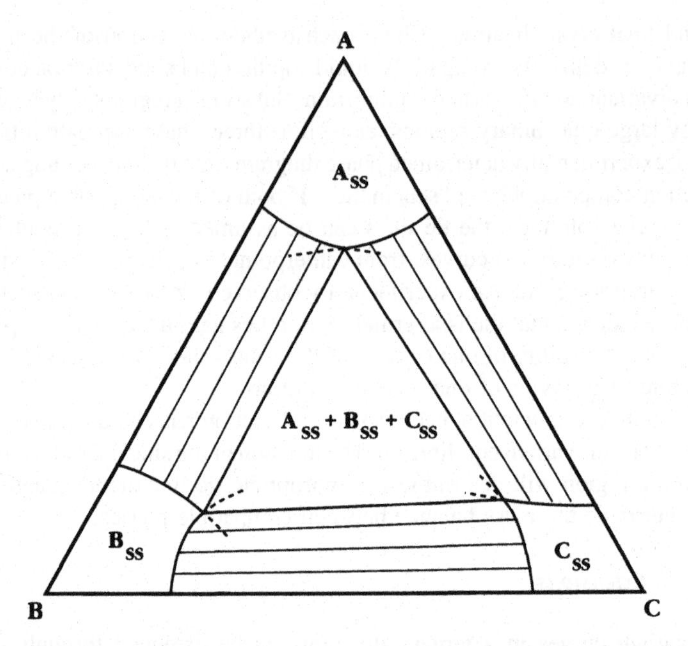

Figure 7.2 Schematic system A–B–C, subsolidus isothermal diagram. A_{ss} and C_{ss} are stable, as indicated by the path of the metastable extensions of the saturation surfaces (both either in the two-phase or in the three-phase regions). This indicates a smoothly varying free energy–composition function for these solid solutions. B_{ss} shows a tendency for instability, most likely phase separation into A-rich and C-rich structure types.

compatibility. While the lever rule is simply applied for two-phase compatibility, three-phase compatibility in a tie triangle requires some added manipulation.

Lever Rule Calculations in the Ternary

Calculation of phase proportions in a three-phase region, or tie triangle, is a matter of conservation of the total system mass as it is distributed among the three compatible phases. Mathematically, the determinant can be set up in appropriate form and solved explicitly for the fractions of phases present. However, it is instructive to explore the properties of the compatibility triangles themselves. The geometric properties relate particularly to the development of families of microstructures in a ceramic system.

Consider the state of system O in Figure 7.3, a diagram of subsolidus compatibility. The composition point O lies in the compatibility triangle B–AC–BC. Point O is seen to lie on the intersection of six lines in either construction (a) or (b); each construction describes related microstructures.

1. In Figure 7.3(a), all lines are constructed from each "apex" or compatible

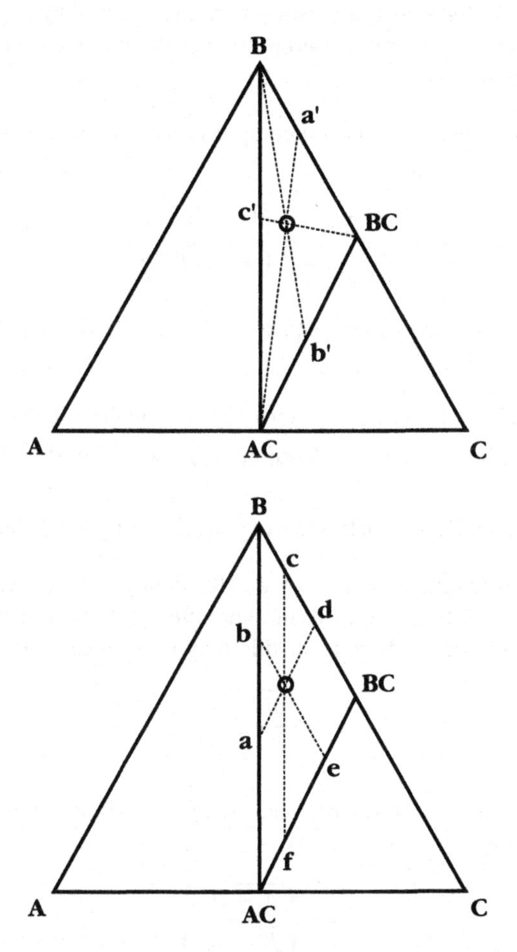

Figure 7.3 Phase analysis constructions for composition O in system A–B–C lying in compatibility triangle B–AC–BC. (See discussion in text.) (a) Demonstration of construction to determine proportion of each "apex" phase. (b) Demonstration of construction of "similar triangles" for phase proportion of each "apex" phase.

phase composition through the composition point to intersect a point of the opposite face of the triangle:

line B–O–b′. All system compositions lying on this line will have phases BC and AC present in the same relative proportion as determined by the lever rule (point b′ is the fulcrum on line AC–BC).

line AC–O–a′. All system compositions on this line will have phase B and BC present in the same relative proportion as determined by the lever rule (point a′ is the fulcrum on line B–BC).

line BC–O–c′. All system compositions on this line will have phases B and AC present in the same relative proportion as determined by the lever rule (point c′ is the fulcrum on line B–AC).

As an example, consider the line of compositions B–O–b′. For all compositions on this line,

$$\frac{f_{AC}}{f_{AC} + f_{BC}} = \frac{(\overline{b' - BC})}{(\overline{AC - BC})}$$

2. In Figure 7.3(b), lines have been constructed parallel to each face through the composition point:

line a–O–d. A line of constant fraction of phase B in the total phase assembly.

line b–O–e. A line of constant fraction of phase AC in the total phase assembly.

line c–O–f. A line of constant fraction of phase BC in the total phase assembly.

Consider the line of compositions a–O–d. All compositions will contain the same fraction of the phase B. At point d, the line intersects the join B–BC, so no A crystals may be present and the lever rule may be applied as in a quasibinary,

$$f_B = \frac{(\overline{d - BC})}{(\overline{B - BC})}$$

It will be seen that the lever rule can be applied similarly at point a on the join B–AC, where

$$f_B = \frac{(\overline{a - BC})}{(\overline{B - AC})}$$

By the geometric rule of similar triangles, the proportion may be taken on any arbitrary line radiating from the phase B composition point, using its intersection with line a–O–d as the fulcrum.

The geometric relationship of these lines to the fulcrum of the mechanical lever is simple. For a physical triangle, in the first instance of Figure 7.3(a), the lines of constant proportion are lines of balance, analogous to balancing the triangle on a fulcrum edge constrained to intersect one apex. In the second instance of Figure 7.3(b), the lines are also lines of balance of the same figure, where the fulcrum edge is constrained to be parallel to one side.

For this example, each line describes a systematic variation in microstructure and properties of the phase assemblies which may be obtained within a subsolidus compatibility triangle. However, the more common application of this geometrical analysis is for univariant crystallization from the liquid, which may be performed for several isothermal conditions.

In the case of univariant equilibrium liquid crystallization, the necessary construction is from the liquid composition point (on the boundary), through the system composition point to the ruled surface or Alkemade line of the compatible solid phases. If this constructed line does not intersect the tie line of the same two phases whose primary crystallization regions are separated by the univariant boundary in question, stop and reconsider your understanding of the isothermal section. Construction of that particular line acts a simple check on the mechanical process of determining the mass balance; once it is determined that the correct compatibility triangle is under consideration, the construction that seems likely to yield the greatest accuracy is preferred.

The method demonstrated below allows moderately simple and accurate calculation of the phase assembly through a combination of line constructions and analysis. The method works with triangles of any proportion or size. Lever lengths are taken in arbitrary units, generally millimeters for convenience, as opposed to the use of compositional units in a binary construction, owing to the random orientation of the compatibility triangle.

Example of Technique

To begin, a line is constructed from one apex of the compatibility triangle through the system composition to the opposite face. Any apex may be chosen, but to minimize error, it is useful to select that which will yield a line most perpendicular to the opposite face. (If the compatibility triangle is very narrow or small, error is impossible to control in any graphical process, and mathematical solution of phase proportion will be necessary using the mass balance equations.)

The proportion of the "apex" phase is determined first, using the lever rule with the system composition as fulcrum. The sum of the remaining two phases is known then by difference and their relative proportion is then determined by application of the lever rule to the opposite "face," or tie line between the two compatible phases. The process may be applied using all three corners as "apex" phase in succession and the result averaged if greater precision is desired.

This technique is preferable to the construction of the parallel lines of constant phase fraction [as in Figure 7.3(b), the so-called method of "similar triangles"], as no construction of perpendiculars is required. The recommended technique is also adaptable to a triangular diagram that has been digitized, with x, y coordinates determined for compatible phases.

Crystallization Mapping in the Ternary

Analysis of crystallization for any single system composition is much easier if the system diagram is first analyzed for regions of common crystallization types. For complicated system diagrams, one with many peritectic reactions, the construction of a mapping of crystallization types is very useful. The crystallization map defines compositional regions in which only one crystallization need be determined exactly to know the salient points describing all possible crystallization paths of compositions within that segment or crystallization type, aiding in the prediction and interpretation of effects of systematic compositional change on phase development and microstructure.

For ceramics that have been crystallized from the melt or sintered with a moderate amount of liquid phase, each of the crystallization types has a distinctly different microstructural appearance and physical performance characteristics. Within each type, a group of similar products may be obtained with some scope for property design. A seemingly minor change in chemistry that results in a major microstructural or performance alteration of a material probably "crossed the line" between neighboring crystallization types.

Solid-state reaction paths and final microstructures obtained are also related to the characteristic crystallization-type behaviors of a system. The Alkemade triangles are the most important predictors of solid-state reaction outcome. In some cases, however, the onset and type of initial reaction may vary in a manner that corresponds to the stable or metastable critical points of the liquidus, reflecting the broader characteristics of the crystallization map.

Crystallization maps can also be the basis for the generation of hypothetical microstructures. Hypothetical problems in equilibrium crystallization can prepare an engineer for the interpretation of microstructures obtained with liquid phase sintering or to extract liquidus data from those sometimes costly mistakes that lead to unexpected melting. Engineering processes do not generally result in an equilibrium outcome, and even exercises in metastable crystallization cannot replace experience in microstructural interpretation.

Constructing the Ternary Crystallization Map

The ternary system's crystallization types are determined using the Alkemade lines, the univariant boundaries, and the invariant points to divide the diagram. It is not necessary to understand all or any of the details of crystallization to construct a meaningful crystallization map from a liquidus diagram containing one or more "triple point" reactions. However, the student or engineer can achieve greater understanding of the processes of crystallization and reaction from using the map construction as a learning aid.

The completed crystallization map divides the diagram into numerous triangles, wedges, and truncated shapes. Each corresponds to a crystallization type with commonality in primary, secondary, and final crystal formation as well as any reaction processes on cooling. The number of unique compositions is infinite, but have now been reduced to a finite number of related sets through the mapping process.

The first step is to construct all Alkemade lines. Many published diagrams include the Alkemade lines. An Alkemade line should be found or constructed to correspond to each univariant boundary curve of the liquidus. The Alkemade lines form the compatibility triangles that define the final solid-state compatibility in each region. Again, note that Alkemade lines will never cross. It is useful to compare the liquidus triple points with the compatibility triangles constructed at this point and to apply the Alkemade theorem to ensure good understanding of the invariant reactions.

The second construction is of a tie line connecting each solid-phase composition to each and any triple point, eutectic or reaction point, for which that solid phase is part of the four-phase invariance described. These lines define the lowest temperature equilibrium solid–solid–liquid tie triangles that can be defined. Each tie triangle is divided further by the univariant reaction curves within the area of the triangle.

Each eutectic will be the centerpoint of a set of three nonoverlapping tie triangles, whereas each peritectic point is the only common point among three overlapping tie triangles corresponding to each univariant boundary entering the reaction point. The tie triangles will each have one side that is an Alkemade line connecting two crystalline compositions. Note that while Alkemade lines cannot cross, the tie triangles may overlap, defining unique regions.

The diagram is now divided into wedgelike composition regions.[3] Within each region, all compositions will share the same sequence of reactions on cooling from the homogeneous liquid phase. Any composition lying on an Alkemade line will have a unique quasibinary or pseudobinary nature, whereas a composition lying on a constructed tie line will exhibit a unique microstructure dominated by the eutectic or peritectic reaction of final crystallization from the liquid.

It is possible to count the number of crystallization types possible in a diagram based on the topological construction of alkemade lines, tie lines, and univariant curves.

sum:
1. Number of triple points $\times$ 6 = ______
2. Number of tie lines that cross a boundary curve = ______
3. Number of tie lines that cross an Alkemade line = ______
4. Number of Alkemade lines that cross more than one boundary curve = ______

The utility of this summation[4] is in ensuring that no unique crystallization behavior is overlooked, as some regions are quite small.

---

[3]A student may find it useful to shade each region in color, so that the mapping will not be obscured in later constructions for calculation of phase proportions.

[4]The summation technique given has been developed by the author through trial and error. Any mathematical proof would be a worthwhile exercise to undertake, as such a proven algorithm could be incorporated into a computerized analysis of liquidus crystallization paths.

Each wedge lies within a primary crystallization region (or liquidus saturation surface), and within a compatibility triangle. If the primary phase and secondary phase are two of the compatible phases, crystallization is simply described.

For a composition lying in a region such as #1 in Figure 7.4 in the primary crystallization region of phase C, a tie line may be drawn from the composition of the crystallizing phase through the system composition to intersect the univariant boundary with phase AC at point x. Such a tie line from any composition point within this mapped area will similarly intersect the univariant boundary curve with AC. Recall that the composition point is the intersection of the system isopleth with the liquidus and indicates the temperature of the onset of precipitation of C crystals. The tie line segment, from the system composition to the univariant boundary, is the first part of the crystallization path, as it tracks the composition of the remaining liquids as only C is removed from the solution (the proportion of A to B in the liquid remains constant on this line; see discussion of Figure 7.3). The lever rule can be applied at any temperature–composition point along this line segment, similarly as it is applied for crystallization from a binary liquid.

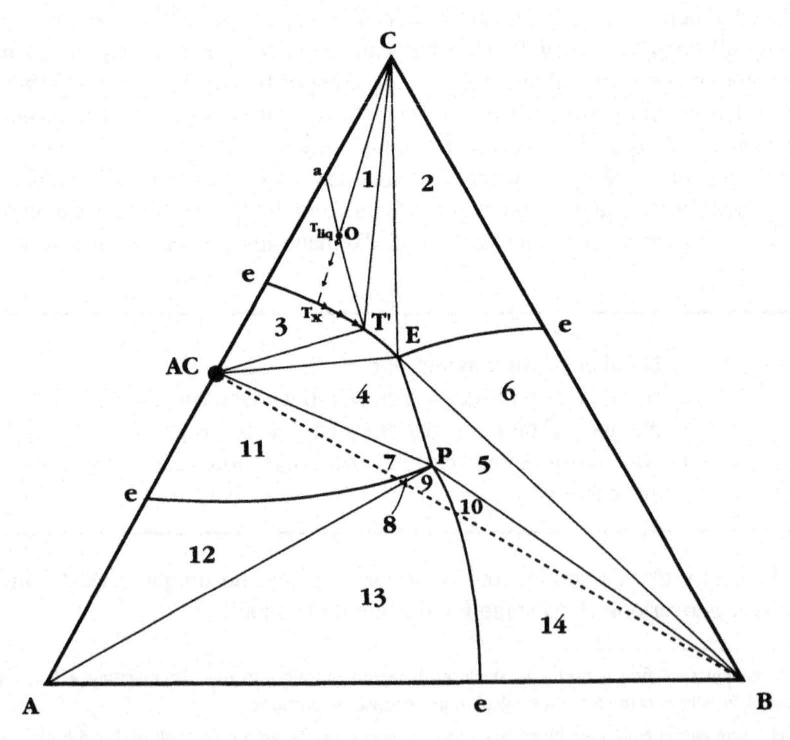

Figure 7.4 Example of determination of common crystallization types with detail of crystallization as described in the text.

The point of intersection with the univariant curve gives both the composition of the "first" liquid that is compatible with both C and AC and also the temperature at which cocrystallization will begin. The intersection of the tie line from C to the eutectic point, a point that is also on the univariant curve, gives the composition and temperature of the last liquid compatible with the two crystalline phases. On any further temperature reduction, the eutectic crystallization of the remaining liquid will occur.

At temperatures between x and E, the point of intersection of the liquidus isotherm with the univariant curve is the composition of the liquid that is in equilibrium with the two compatible phases described by the ruled surface. The phase proportions in the triangle liquid–solid C–solid AC are determined in the conventional manner.

At the temperature of the eutectic, only the three phase equilibria just above and just below the eutectic temperature may be described exactly. Below the eutectic temperature, the proportions of the three compatible solids are determined from the location of the system composition point in the compatibility triangle.

Continuing with Figure 7.4; crystallizations in regions 1 to 6 are all simply described similarly as for region 1.

Regions 7 to 10 do not satisfy the criteria for simple crystallization as both primary and secondary crystalline phases are not two of the three phases forming the compatibility triangle. Regions 7 to 10 lie in the compatibility triangle AC–C–B, but all involve crystallization of phase A as either primary or secondary crystal and the reaction of crystalline A with the peritectic liquid to form AC + B.

Note the relative size of region 8, very easily overlooked in cursory examination of the system. Although small, this area of system compositions could have very useful microstructure depending upon the relationship between A crystals and the nucleation of the AC phase. In this example, nonequilibrium crystallization of B could not occur prior to the temperature of the peritectic, as any tangent to the univariant boundary A + AC + liq intersects the join A–AC, indicating cocrystallization.

Regions 11 to 14 are simple crystallizations, although their final reaction is described by the peritectic.

In the second example (Figure 7.5) the mapping again yields 14 regions. Regions 1 to 3 are simple crystallizations that end in the peritectic reaction.

Regions 9 to 14 are simple crystallizations ending with eutectic crystallization.

In regions 4 and 5, crystals of C are formed that will be consumed in the peritectic reaction, C + liq = C_2B.

In regions 6, 7, and 8, all in the primary crystallization region of C, the primary crystals will react with the liquids along the reaction boundary. In 6, sufficient C is present so that crystalline C will be available for reaction at all temperatures between the system liquidus and the peritectic reaction temperature. In regions 7 and 8, however, all primary *C* crystals will be consumed by reaction to form C_2B prior to the peritectic temperature. When the last C crystal is reacted, the

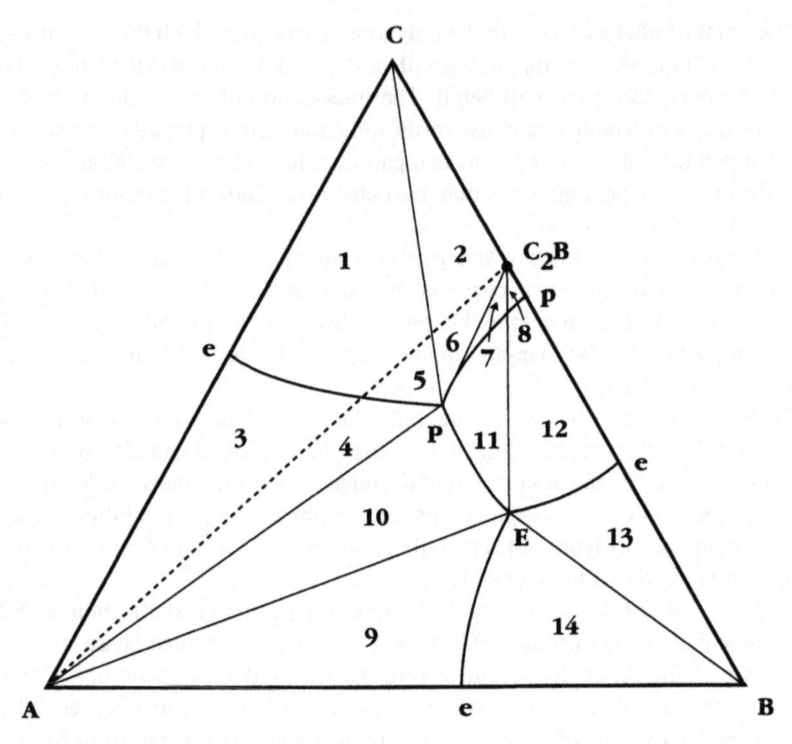

Figure 7.5 Example of construction of diagram of common crystallization types.

system is again divariant, liquid + C_2B, and the crystallization path will follow the path of the tie line from C_2B, through the system composition to intersect either the univariant boundary with A or B for regions 7 and 8, respectively. Again, it is the property of triangles that the tie line from the crystallizing phase composition, C_2B, through the system composition point is a line of mass balance, where the proportions of A to B in the liquid are consistent with mass balance as C and B are removed in stoichiometric proportion to form the compound.

7.3 Special Ternary Diagram Presentations

The equilateral triangle composition plane is traditional in oxide systems, but other formats are as readily interpreted. The variant most generally seen is a right triangle or square, having rectilinear composition axes, seen in "reciprocal salt" and water–salt diagrams.

For condensed systems in which two complex anions are present, a reciprocal salt diagram is often used. An example is given in Figure 7.6. Each axis represents a progressive substitution or exchange reaction. For many reciprocal systems,

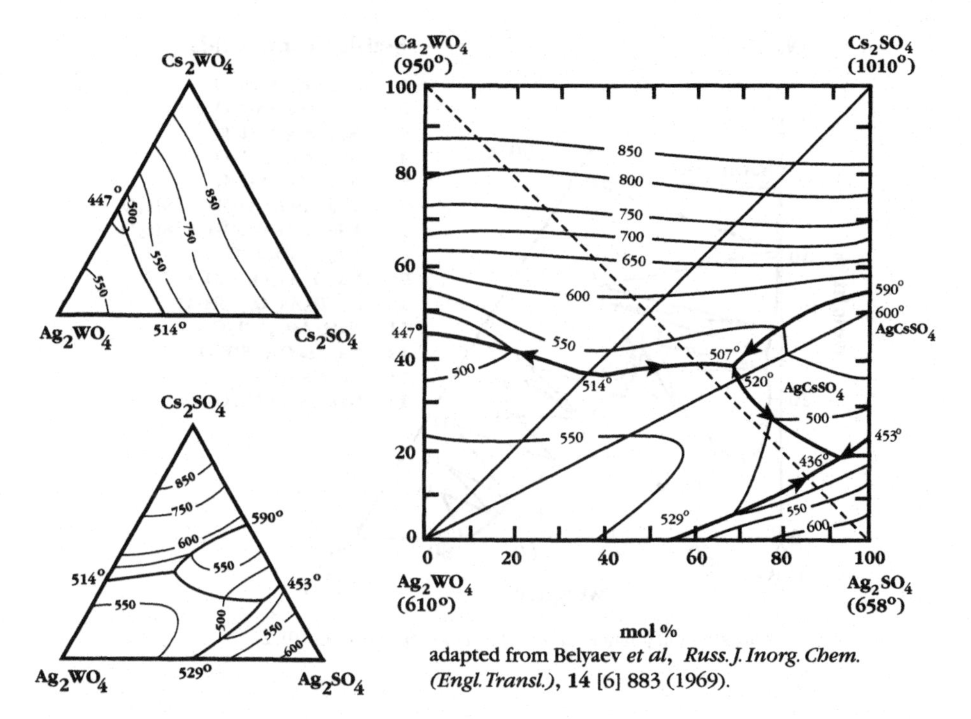

Figure 7.6 An example of the reciprocal salt diagram presentation with inset diagram of alternate ternary styles.

an alternate setting is as a set of two ternary diagrams, if the diagonal is a true binary. Each side of the diagram is a quasibinary system.

A single-system composition is located using the rectilinear axis units or as a point in a composition triangle, if appropriate for the starting phases. A crystallization map can be derived for the reciprocal salt system using similar procedures as described above, and mass balance of compatible phases is determined using the lever or balance rules as required.

The reciprocal salt presentation is particularly useful in systems where substitution of anion groups and the formation of complex compounds is possible. The "best" presentation for a particular system may vary with the industry or research application. For instance, in systems of interest in the hydration of portland cement, the presentation should unambiguously represent known metastable or disequilibrium reactions.

The equilateral right triangular presentation of Figure 7.7 is quite commonly seen in water–salt systems. This type of system, although presented as if it were ternary, is truly quaternary if the occurrence of hydrated salts (containing OH^-) is noted. The significant data presented is the saturated solution concentration at a fixed temperature. The lever rule may be applied.

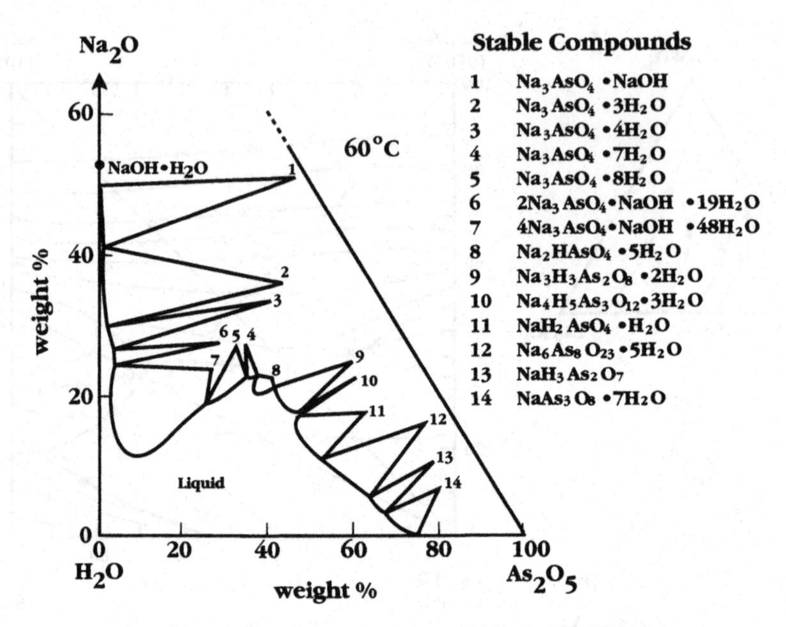

adapted from Jouini *et al*, *Bull. Soc. chim. Fr.*, No.1 ,66 (1972).

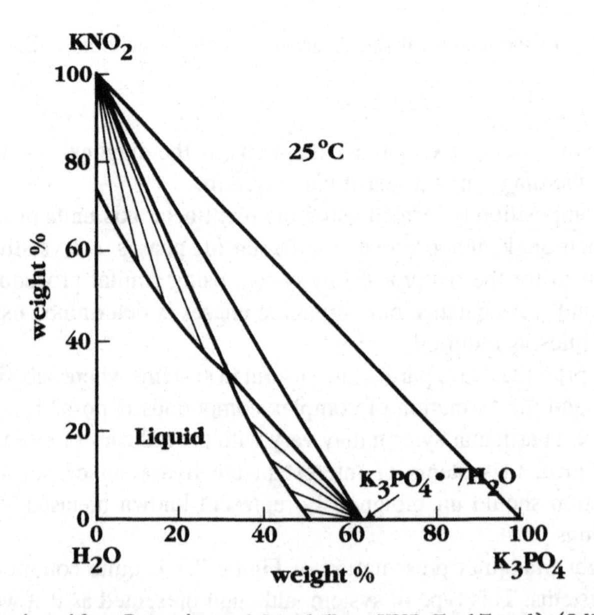

adapted from Protsecko *et al*, *J. Appl. Chem. USSR (Engl. Transl.)*, **48** [5] 1104 (1975).

Figure 7.7 Typical examples of water–salt diagrams.

Other significant physical property data may be collected into a "phase diagram." A property diagram is particularly useful for the secondary characteristics of extensive solid solution phases, both equilibrium and kinetically stranded or "quenched" phases. For equilibrium phases, such as complex dielectrics used in electronic and magnetic applications, mapping of second-order properties can indicate regions of stable and unstable behavior. In microwave dielectric ceramics, minimal dielectric loss corresponds to the greatest solid solution or compound stability. Relative phase stability is difficult to determine using conventional solid-state annealing, and may be more apparent as properties are systematically plotted.

The physical properties of glasses, density and refractive index, are commonly presented as is their relative stability against crystallization, as in the T–T–T curve. Stability diagrams such as the T–T–T curve are also used in the annealing of metals and in the presentation of "aging" data in complex crystalline oxide solid solution phases, for refractory or electronic applications.

7.4 Quaternary and Higher System Diagrams

Any two-dimensional presentation of phase equilibrium can only present effectively the restricted conditions which have a variance of 0, 1, or 2 according to the phase rule. For the one-component system, no assumptions were required ($\Phi = C - P + 2$). For two-component systems, the condensed, isobaric system could be effectively presented ($\Phi_c = C - P + 1$).

Ternary systems of condensed phases are shown for isobaric conditions also, and form a three-dimensional solid figure with a triangular composition grid as base and temperature as the vertical scale. To present di- or univariant phase equilibria as dependent on both temperature and composition in a single ternary diagram (or projection), only the solution phase composition and temperature are required to define the state of the system through mass balance, as all other phase compositions are assumed constant. These assumptions are extended to allow presentation of simple crystallization in an isoconcentration plane of a quaternary system.

The two-dimensional diagram features—point, line, and area—represent the system variances of 0, 1, and 2. These correspond to five-, four-, and three-phase compatibility in a four-component system.

A quaternary "liquidus diagram" can be represented reasonably accurately in a triangular isothermal composition plane for the case where one component is significantly more refractory or higher melting than all others in the system, and also has near zero solubility of or in the other components. On cooling, this refractory component is the first to crystallize from the quaternary melt for all compositions. The "liquidus" surface then presented is the surface of *secondary* saturation, and the boundary curves are intersections of these secondary surfaces. The apparent "system" composition is made up of the remaining three components. This presentation technique is very commonly used in systems where a constant fraction MgO [T_m(MgO) ≈ 2800°C] is maintained, but could be useful in other systems.

Each constant-fraction plane is presented as a unique member of a set of related phase diagrams. An apparent "triple point" is not necessarily an eutectic (absolute liquidus minima) or peritectic point, however, and is termed a "piercing point." A "piercing point" is a liquidus invariance for five phases (liquid + four solids) found where the activity of the refractory component has the fixed value of the isoconcentration plane. The three secondary saturation surfaces meet at a different liquid composition point and temperature if a different isoconcentration plane is examined. The true eutectic point would be the absolute minimum temperature below which no liquid could be observed.

If it can be assumed that essentially all of the refractory component is contained in the crystals present, the lever rule and other constructions applicable to a ternary liquidus diagram may be cautiously applied in those regions of the quaternary diagram in which a figure of mass balance (a triangle of compatible phase compositions) can be imagined. The secondary saturation surfaces and their intersections are valid thermodynamic constructions and have valid metastable extensions. Special care must be taken to ensure conservation of mass in phase equilibria calculation in any quaternary system, keeping in mind the phase rule.

A similar triangular planar presentation could be used if one component was consistently the last to crystallize and did not appear in significant solid solution in the other solid phases, as a flux used in crystal growth. The discussion above as to the difference between piercing points and true invariant points still applies. The form of the diagram is dependent on the presence and concentration of the fourth component in the liquid phase. However, now the "liquidus" diagram would represent the surface of primary saturation of the liquid phase. The true solidus (occurrence of four compatible solids with a critical liquid) is located at a far lower temperature than represented in the diagram.

A reasonable and common presentation of four- and five- and sometimes six-component equilibria is found in isothermal section presentation of concentration planes formed by important mineral phases. Such a plane is a cut of a complex prism and may not be trigonal in shape. The only requirement that may be made is that all compositions of all phase represented be composed of the given endmembers. Thus a liquid or solid solution composition may be determined exactly using the same center-of-gravity analogy as used for a triangle, but applied to the more complex shape. The lever rule may still be applied in two phase areas.

A small problem arises in interpretation of the presentation if the scaling of the plane is truly that of the overall prism from which it has been extracted, particularly if the prism layout is in terms of mol% prism components. A wt% scaled diagram can usually be interpreted unambiguously.

Clear labeling of all axes is essential for manipulation of any multicomponent phase diagram data. In systems with greater than four components, diagrammatic

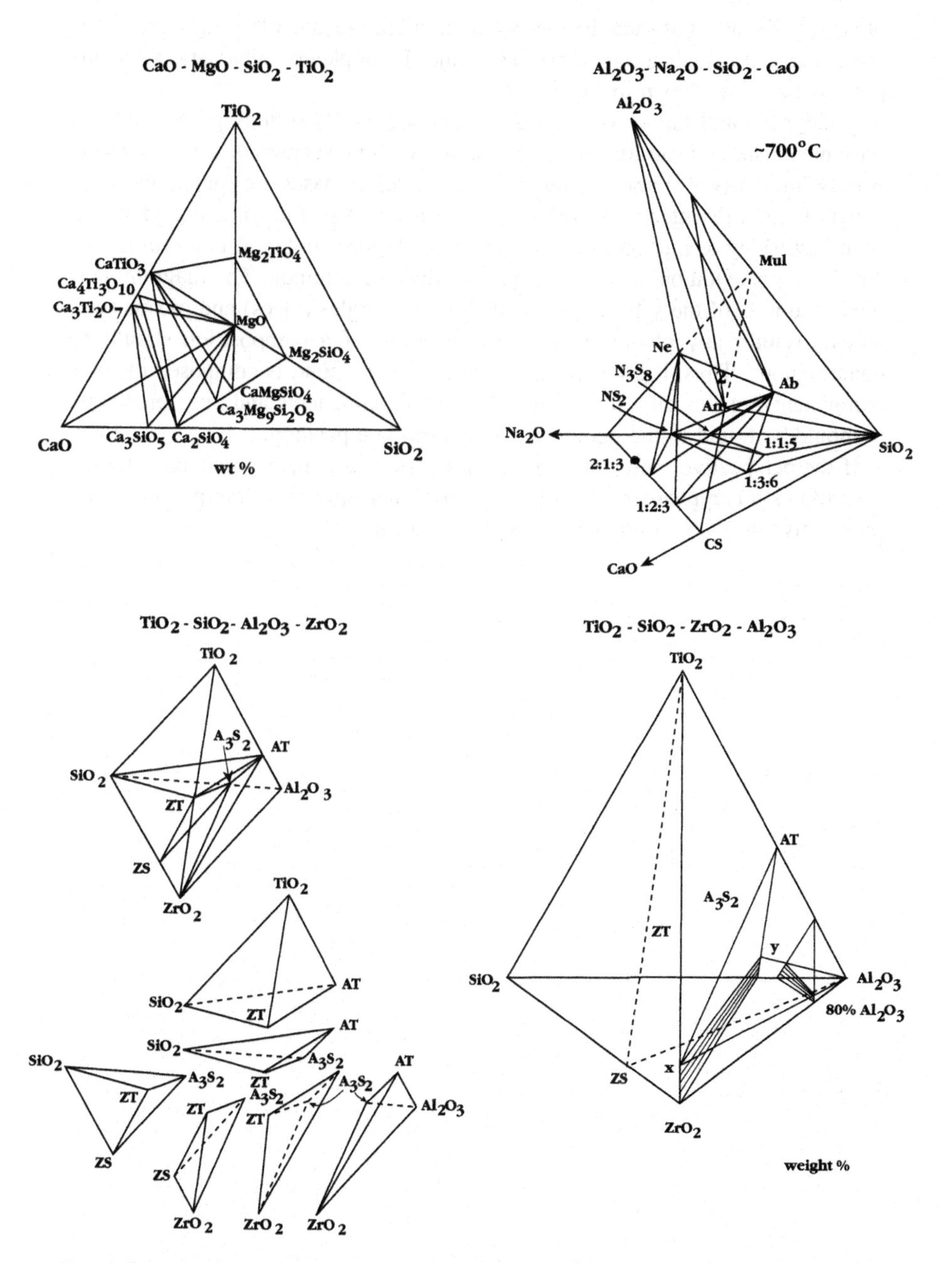

Figure 7.8 Schematic multicomponent prism presentations. Not all systems can be presented unambiguously in this manner.

presentations as published in the research literature are often schematic and not suitable for application of the lever rule. Examples of schematic prismatic presentations are shown in Figure 7.8.

Quaternary and higher order (up to six component) solidus prisms may be presented also, and can be very informative. Such presentations are most useful if only limited (<5%) solid solution ranges in solid phases are found, and most generally are presented to accompany more detailed sections mentioned above.

In first trying to interpret the quaternary or "higher" prism, it is useful to first "test" the presentation, as errors in perspective are common. The diagrammatic presentation is of the subsolidus compatibility triangles (faces) and prisms (volumes). Within any planar triangle, the location of a composition should be unambiguous. Thus, if a ternary or quaternary compound is composed of three components, or lies on a plane formed by two components and one compound, it should be found in the geometrically appropriate position.

If the perspective seems true on various planes, it is then possible to use the prism to estimate phase proportions of subsolidus equilibria. Interpretation of a perspective drawing is not trivial for most viewers.

8

Engineered Systems and Oxide Phase Equilibria

Oxides are uniquely stable against physical and chemical interactions. Their stability against reaction has been discussed in terms of their inherent stoichiometric constraints on defect formation and atomic transport which limit the progress of reaction. Their complex phase stoichiometry is a statistical limit to the probability of equilibrium phase formation and can restrict the progress of even nominally homogeneous phase transformation. The equilibrium characteristics of each reacting system predicts the long-term outcome and the engineer must design each stage of the thermal and chemical processing to minimize constraint on desirable reaction while maximizing the stability of the final product. Engineers attempt to manipulate local kinetic and chemical factors to promote a short-term result with useful ceramic properties.

The many techniques of characterization—chemical, structural, and microstructural—are used to determine the present state of the system and its phases. The engineer must also promote stability in the product. Ceramic oxide systems are most generally observed in the nonequilibrium state, not only in the many stages of processing but also as products in the application environment. The equilibrium phase diagram presents the reference state of minimum free energy of a chemical or materials system. The engineer determines the relationship of the current state to the reference state and to the desired product state. The current state can be characterized not only by its overall chemistry and phases, but also by the relationship between those phases and their state of equilibration.

An unreacted or mechanical aggregate system is an assembly of phases that have not been exposed to thermal or chemical environments that would promote sufficient atomic motion for significant reaction to proceed. Examples might be a fiber-reinforced plastic, a "green" (or unfired) ceramic body, and a stamped bimetallic junction or metallic contact to a semiconducting or insulating material.

An aggregate system may have been subjected to an active environment sufficient to initiate reaction but not to achieve or approach equilibrium. A porcelain ceramic or conventionally sintered ceramic, a reaction-bonded ceramic or "cermet," an hydraulic-bonded portland cement paste, and a part formed by infiltration of a liquid phase into a porous body or metallurgical shapes formed by powder metallurgical techniques are examples. The limited reaction achieved and the reaction products formed may not represent equilibrium phases of the overall system.

Systems that have been equilibrated with an environment at one time but are not at equilibrium at the time of observation are considered stranded. Oxide glasses, fusion cast refractories, and molten metal castings are examples. Some sintered ceramics, particularly high-performance ceramics, are of this type, as are most mineralogical phases used as raw materials.

8.1 Conditions and Kinetics of Reaction

Physical change occurs slowly in a closed system of solid oxides owing to high activation energies and the necessity for correlated defect formation and atomic transport. Reaction processes and transport kinetics in all of the available pathways must be considered for the effective correlation of observed physical systems with their equilibrium states. The observed state of an oxide system will be that which was both energetically favorable and kinetically achievable and will often reflect the multiple reaction processes from initial to final condition.

When two pure dissimilar oxides are brought together in intimate contact, the chemical potential functions are discontinuous across the interface, with slope of $\pm\infty$ at the point of contact. Any surface transfer of species or reaction immediately reduces the gradient to a finite value. Thus, initial reaction is a spontaneous and irreversible change of the system. While initial reaction is always to a lower energy state, the product is not necessarily one of the phases that would be present under equilibrium conditions.

In an isothermal system, a coherent surface reaction product is a physical response to the thermodynamic requirement for a minimization in local chemical potential gradient. Such reaction will take place spontaneously as soon as sufficient energy is available. Our normal measure of available energy is absolute temperature, an average over the statistically large number of individual particle states in a large volume. Loss response to electromagnetic fields and mechanical loss from acoustic response contribute to locally high "effective" temperatures for resonant states. Surface frictional heating can also be an important energy source for surface reaction. Local temperature fluctuations are statistical events seen to promote local reaction processes.

The phase diagram presents an isothermal closed system under conditions of reversible equilibrium without kinetic restraint on that reversibility. It is not suitable to use a phase diagram to predict the course of the initial, irreversible, reaction in a nonequilibrium system. The initiation temperature of the reaction may be estimated from the equilibrium data, however, using melting point and eutectic data from the diagram or diagrams of the system.

Initial reaction, being predicated on an availability of reactants and the relaxation of volume and stoichiometry constraints, is a surface phenomenon. As the surface is altered by reaction, the reaction's nature and kinetics are changed. For many engineering purposes, sluggish and kinetically limited surface reaction behavior is beneficial in applications that require stability. Sluggish reaction kinetics are responsible for the practical application of corrosion-resistant or passivation coatings on bulk surfaces.

The state of that product of fastest reaction, its location or morphology, influences the course of slower processes. If the initial reaction product is a liquid, it may dissolve other reactants and promote overall reaction. A liquid may also serve as a growth medium—fast transport through the liquid could promote dendritic or columnar growth. Those high aspect forms do not efficiently fill the void space and also tend to physically separate slower reacting phases. On the other hand, if the fastest reaction formed a compact solid layer, its influence on succeeding reaction or product morphology might be through the isolation of one component from the other reactants. Reaction to isolate and limit reaction is desirable in some composites, for instance, but can be very frustrating when a multicomponent single-phase product is desired from solid-state reaction.

Kinetic processes that limit or control initial or secondary reaction progress and can be externally controlled (through time, temperature, initial microstructure, defect or surface chemistries) are often used to advantage in the fabrication of unique microstructures. These would include internal boundary layer devices, reinforced composites, and in situ composites. These products require limited surface reaction or controlled surface segregation during fabrication.

Reaction processes are separated here from nucleation phenomena, which will be discussed in a later section. Although nucleation is not generally a limit to the initial reaction, nucleation kinetics may be the deciding factor in the appearance of a metastable versus equilibrium reaction product.

A Simplified Reacting System

The system under consideration may be closed, having all nonvolatile components, or open and contain one or more mobile or volatile species. For a closed or nonvolatile system, initial reaction is confined to contact areas between reactants. The typical assembly of phases will contain a variety of interface types, as schematically diagrammed in Figure 8.1. Although first reference to the ternary system phase diagram indicates that at equilibrium only the single phase ABC is present, no three-phase junctions exist in the initial assembly. Binary interface reaction must physically precede ternary or higher reaction.

The homologous temperature, $T_h = T/T_m$, of a phase is a measure of the available energy to approach structural equilibrium. Similarly, in a multiphase assemblage

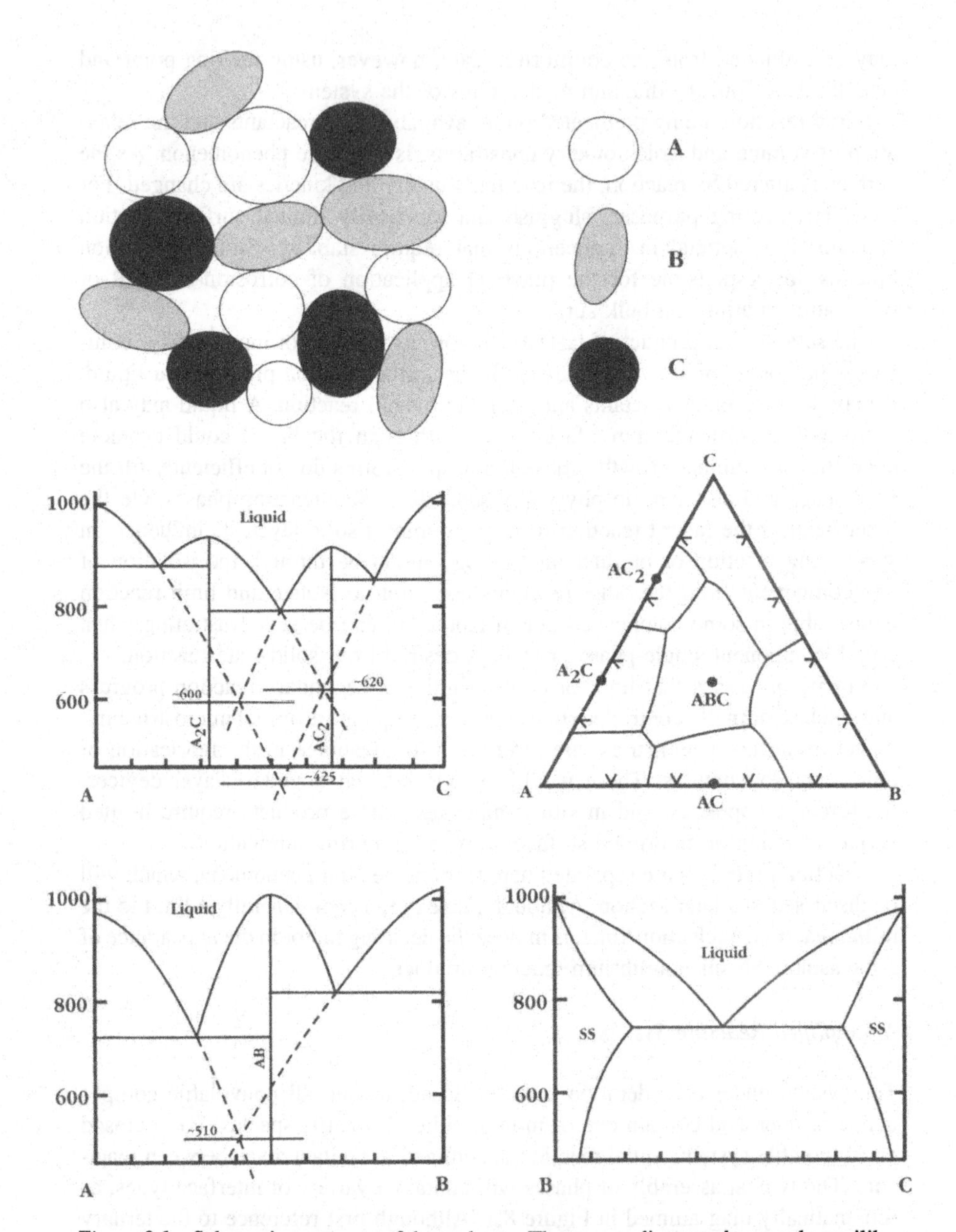

Figure 8.1 Schematic aggregate of three phases. The ternary diagram shows only equilibrium data. Probable pairwise interactions are determined through reference to the appropriate binary phase diagrams.

of reactants, the lowest eutectic temperature of any pair is the highest temperature at which the original phases of a nonequilibrium system can remain unreacted. It is often the case that reaction will be observed at 80 to 90% of this temperature, although the rate of reaction may be very slow unless a metastable liquid can form. Rapid reaction in a nonequilibrium system depends on very fast transport of all reactants that cannot occur in the solid state below about $T_h = 0.8$ for the most refractory phase.

The binary diagrams indicate three possible equilibrium phases as candidates for "first-to-form"—there may be many metastable products that are not represented on these equilibrium diagrams—and that reaction should be observed first between either A–B or A–C at a temperature of about 520 to 620°C (0.8 of the lowest eutectics), which give a "window" for investigation. A practical problem in process design for solid-state reaction is theorizing on the nature of the first product phase and the temperature of its formation, and a serious question is "of the three (or more) possible phases, which is most likely to be first?" Once the most likely system for first reaction is determined, and the nature of the reaction product, the best means of its detection and the possible influence on further reaction can be discussed and theories meaningfully tested.

Cation Characteristics and Reaction

Reaction between oxides is most favorable when the cations are of highly different electronegativity or ionization potentials. Both are measures of the relative energy to ionize the cation; the electronegativity measures the relative strength of the cation to attract shared bonding electrons while the ionization potential is the energy required to totally remove an electron from an isolated cation.

Electronegativity and Ionicity

Cation electronegativity is predictive of the relative ionic character of the metal–oxygen bond. A difference of about two units (anion minus cation electronegativities) indicates predominantly ionic bonding. Most common oxides are in the range 1.5 to 2.6, or 30 to 80% ionic character in oxide bond formation. A trend in values of elemental electronegativities can be seen as increasing from left to right across the periodic table with some irregularities in the transition metals. A large electronegativity difference in oxide formation indicates a very stable directional bond and correlates with high melting point. The higher electronegativity of oxygen relative to carbon or nitrogen is reflected in the greater stability of the metal oxide with respect to the carbide or nitride of the same metal. These compounds, and also the metal borides, will tend to oxidize at the surface. The oxide is generally passivating if it remains coherent, and limits any further reaction until high temperatures. (See "tarnishing".)

Ionization Potential and Acid–Base Behavior

Cation ionization potential is used to form the basis for an evaluation of the expected acid–base behavior of oxides. Acidic or basic character of oxides is one of the oldest conceptual design schemes for refractory compatibility and corrosion resistance. The basic oxides (MgO, CaO) are generally found to the left of the periodic table, while acid oxides (SiO_2, P_2O_5) are found at the right.[1] In common analogy with simple aqueous chemistry, an acid oxide tends to react with a basic oxide to form a reaction or corrosion product. Alumina and zirconia tend to be amphoteric, reacting to form products with either acidic or basic oxides. Thus, a basic refractory such as magnesia is not suitable for extended contact with an acid (silica containing) slag or melt, but mag-chrome refractories (more amphoteric) may show little or no reaction.

The more general measure of acid or base nature of oxides is through estimation of the pH_0, or "point of zero charge of the immersed solid," which is estimated using the cation ionization potential. The simple relation

$$pH_o = 12.22 - 0.89\ (IP)$$

where IP =
total ionization potential (MJ/mol) [electron volts (eV) × 0.09648 = MJ/mol].

results in a scale similar to traditional aqueous values. Estimation of pH_O also enables comparison of the acid/base nature of the oxides of variable valent cations. (The ionization potential used is the full summation to the cation valence of interest.)

Some representative values of pH_O for common oxides are presented in Table 8.1. An acid/base ranking scheme for oxides is usefully applied in evaluation of tendency of reaction in most material systems, including nonoxide ceramics and nonpristine metals, as initial reaction is generally predicated on surface reaction. All but the most pristine surfaces will reflect a degree of oxide formation, a so-called "native oxide." Pristine nitride surfaces tend to be amphoteric while carbides tend toward mild basic behavior.

Acid–base behavior in more complex multicomponent oxide compounds and solid solutions can be explored through the more formal (although still empirical) "bond-valence" methods. These methods are used for the evaluation of the tendency of formation of reaction products and intermediates and the relative stability of these against further alteration. The bond-valence value is based on formal valency of the species, such as from the periodic table, divided over its multiple

[1] The traditional acid or base nature of these oxides is actually relative to observed behavior in aqueous environments under applied potential conditions. See Pourbaix diagram discussion in Chapter 6.

Table 8.1 Representative Values of pH$_o$ Calculated Using Ionization Potential of Cation (Values for IP are from CRC Handbook of Chemistry and Physics.*)*

Oxide		IP (MJ/mol)	pH$_O$	
Li_2O	Li^+	0.569	11.7	SB
Na_2O	Na^+	0.493	11.8	SB
MgO	Mg^{2+}	2.177	10.3	SB
CaO	Ca^{2+}	1.726	10.7	SB
BaO	Ba^{2+}	1.461	10.9	SB
Sc_2O_3	Sc^{3+}	4.234	8.45	WB
ZrO_2	Zr^{4+}	7.421	5.62	WA
CoO	Co^{2+}	2.39	10.1	SB
NiO	Ni^{2+}	2.48	10.0	SB
ZnO	Zn^{2+}	2.63	9.9	SB
B_2O_3	B^{3+}	6.853	6.1	WA
Al_2O_3	Al^{3+}	5.14	7.64	amph
SiO_2	Si^{4+}	9.90	3.4	SA
P_2O_5	P^{5+}	16.898	(off scale)	vSA
PbO_2	Pb^{4+}	9.28	3.95	SA
PbO	Pb^{2+}	2.156	10.3	SB
TiO_2	Ti^{4+}	8.75	4.43	SA
Bi_2O_3	Bi^{3+}	4.75	8.0	WB
FeO	Fe^{2+}	2.32	9.9	SB
Fe_2O_3	Fe^{3+}	5.278	6.94	amph

A = acid; B = base (S = strong; W = weak); amph = amphoteric.

bonds in a manner inversely proportional to the relative bond length. (These values are averaged over data from many stable structures and are tabulated in references cited below.)

In a stable structure, the "bond-valence summation" around each cation or anion is equal to the formal atomic valence. Local change in chemistry and coordination due to interaction can then be evaluated through the change in the bond-valence sum of the postulated intermediate or reaction product. It is a necessary requirement that the local bond-valence sums remain close to the formal values during all stages of a chemical interaction such as solid solution formation or local structure formation prior to phase nucleation if the proposed reaction is to be energetically favorable. A full discussion of bond-valence sum rules and their relation to coordination chemistry is outside the scope of this book, but can be found in Brown (1978) and Brown and Shannon (1973) as well as other sources on experimental crystal chemistry.

Acid–base type ranking can be used to postulate the initial sequence of possible reactions, or the "fastest" initial reaction among many possibilities. The reaction of a strong acid with a strong base is expected to proceed more readily than the reactions of weaker or more amphoteric oxides. Acid–base arguments using pH$_O$

values are best to determine the first phases to react. Further reaction tendencies of more complex phases can be evaluated using the more formal bond-valence summation methods.

Reaction with a Vapor Phase

Reaction with a vapor phase or the volatilization of a reaction product may occur in either a closed or open system condition. The initial surface chemistry and structure of the individual reactant phases will usually reflect some prior interaction with vapor components. The initial reaction of a metal nitride is that of the "native" surface oxide. Similarly, acid oxides have an affinity for hydroxyl groups and basic oxides for sulfate and carbonate species. While the presence of adsorbed species "buffers" the estimated acid or base behaviors at low temperatures, volatilization will usually occur between 100 and 600°C, prior to the onset of initial solid-state reaction between refractory phases.

Volatile species have limited effect on initial or secondary reaction of refractory phases if generated in small quantities and when the vapor is free to escape to the atmosphere. In some situations, such a deep powder bed reactor, the path to the atmosphere is tortuous and volatile species are real components of the system and contribute to reaction processes.

The normal presence of volatile surface adsorbed species in a oxide assembly makes necessary the consideration of tarnishing, chemical exchange, and chemical vapor transport processes on the course of further reaction at higher process temperatures. Tarnishing, a reaction of the type "solid + gas = product," results in the formation of a surface film that may isolate the reactant or moderate its participation in further reaction.

Chemical exchange reaction, on the other hand, of a type AX(s) + BY(g) = AY(g) + BX(s), promotes transfer of cation species via the vapor phase and could lead to the destruction of an initial binary product in favor of equilibrium multicomponent phase formation or the nucleation of an equilibrium phase when no other physical transport path is present between reactants.

Chemical vapor transport reaction results in the redeposition of material, generally at a lower temperature location in a nonisothermal system, but without reaction. This process, in an isothermal system, leads only to shape change or elimination of fine particles. The volatilization of adsorbed halides is usually responsible for chemical vapor transport processes. Redeposition due to chemical vapor transport processes at a low temperature can be responsible for the initiation or modification of later reaction.

In many important traditional ceramic systems, the principal raw materials will be hydroxide, sulfate, or carbonate minerals. Although these systems, which include clay products, portland cements, and traditional refractory grogs, are discussed in terms of oxide products, their initial reactions are much more complex than those of simple oxides as mass is not conserved at the reaction interfaces.

The multistage reaction equilibria in these systems are areas of continuing interest in manufacturing. Because of the presence of volatile components, and the morphological and surface activity changes that occur to the initial mineral phases, initial reaction in these important technological systems cannot be readily predicted using oxide phase diagrams.

8.2 Reaction in Ceramic Processing

Mechanical Aggregate Systems

Aggregate systems generally are mechanically stable assemblies of nonthermodynamically compatible phases. The binding phase may be moderately reactive or interactive with one or more of these phases, or may be simply an adhesive. Adhesive bonds would include the obvious polymeric compounds, such as the resin phase in fiberglass composites and the latex pigment carriers in paints. Hydrates, hydroxides, alcohols, and some polymers are characteristically polar species and interact with the surface defect states of oxides, with the result being a reasonable adhesion and mechanical stability. Nonpolar molecules are not efficient binders but may result in mechanically strong aggregates or composites if the nonpolar phase is continuous and space filling. Water must be considered as moderately reactive and caustics more so, as hydrates are more thermodynamically stable than oxides at ambient temperatures.

Thermodynamic equilibrium is neither required nor desired from most of the engineering aggregates, which are heterogeneous composite systems. Examples include a glass fiber reinforced thermoplastic resin and flexible magnetic recording tape, both having oxide particles that have been aggregated in a polymeric matrix. The polymeric compound is processed with the oxide filler or reinforcement at temperatures far below $T_h \sim 0.2$ of the oxide phase. At such low temperatures, only reaction of surface contaminants or bound hydroxyl groups with the binding phase need be considered.

$T_h < 0.3$ to 0.4 is usually the limit of temperature excursion during composite or heterogeneous device fabrication without consideration of the effects of at least interfacial chemical interaction between the device components, particularly metallic contacts and coatings. This covers a range from about 200 to 500°C (depending on melting point) for the less refractory oxides and glasses. While this is well above the temperature range of oxide–polymer composite processing, it is below the temperature range for oxide–metal composite fabrication. For example, brazing, the formation of ceramic–metal bonds which requires the development of a graded interface for strength, is typically performed at a higher temperature, $T_h \leq 0.5$, of the ceramic. The brazing alloy must be designed to flow in this temperature range to form a continuous coating at the interface. Very short times are used, so that reaction interaction depth is limited to only a few microns.

Ceramic phases and phase mixtures are prepared as aggregates for the reactive processes of calcination and sintering. The engineer designs the aggregate, considering the phases and their relative particle sizes, for reaction at a specific temperature. Time and temperature can be manipulated to promote metastable or stable bulk and surface reactions using the phase diagram and available kinetic data.

Aggregate Design for Reaction

A ceramic system is composed of its base oxides or compounds, and additives. For application of phase equilibrium information in the understanding of reactive processes, the physical distribution and characteristics of the oxide phases must be known. Nonuniformities in particle packing are reflected in non-uniform reaction.

Uniform mixtures are defined in terms of dimensions of the system being observed. A system is statistically random and uniform if a small sample, taken by accepted techniques, has the same composition as the overall mixture. For meaningful random sampling of a mixture of crystalline phases, the sample (linear) dimension is at least 100× the dimension of the largest particle size fraction. For example, if 150 μm was the largest particle size in the aggregate, the smallest representative sample for observation would be on the order of 4 cm^3 (or $(100 \times 150\ \mu m)^3$). This is the smallest part of a well-mixed sample that would be expected to behave as an "infinite" system as portrayed by the phase diagram, necessary for predicting stable or metastable reaction behavior.[2]

Reaction processes initiate locally and may be simple binary, ternary, or more complex in nature, depending on the interparticle contacts in the aggregate. The actual particle sizes and particle size distribution determine the nature and scale of the "system" or "systems" that characterize the as-fabricated, unreacted, microstructure.

The equilibrium phase diagrams will quickly show whether the assembled phases are *pairwise* compatible at any temperature. Regardless of the overall complexity of the aggregate system chemistry, any solid-state reaction must initiate at a point of contact between two phases. (If one or more components are present in the form of a frit or complex mineral phase, a quasibinary reaction in the ternary system is considered.)

A typical aggregate of three oxides is diagrammed in Figure 8.1.

The pairwise diagrams, A–C, B–C and A–C, indicate that this assembly is not thermodynamically stable and that reaction can be expected among the phases. The equilibrium phase diagram does not indicate phase formation onset tempera-

[2]Good practice in raw materials selection reflects these dimensional requirements, although not usually stated as statistical requirements for reaction but due to strength and soundness requirements in the finished ware. For a ceramic body with a wall thickness of about 6 mm (defining the linear dimension of the sample under observation), to show reasonable uniformity for reactive sintering, the maximum particle size fraction desirable for reaction uniformity is about 60 μm. Coarser particulate phases, generally nonreactive, may be incorporated into the mix to control shrinkage.

tures, only equilibrium data. However, through careful construction of the metastable extension of the equilibrium liquidus curves for the phases present, diagrams of metastable stability can be sketched as indicated.

The liquidus curves and the solidus curves of a solid solution in binary phase diagrams can be extended as these are true projections from the minimum free energy surface onto the composition plane. In diagrams that are ternary and higher, metastable extension of similar univariant curves may be performed, but with caution as subsolidus phase composition and reaction cannot be represented unambiguously.

The intersection of the metastable liquidus curves represents the maximum temperature at which the two solids in question can exist in true metastable equilibrium without reaction in the bulk state. High surface energy phases (highly defective or damaged surface owing to grinding or crushing, or defective owing to impurities adsorbed or incorporated) may be surface reactive at significantly lower temperatures.

As the accuracy of the graphical estimation for this temperature is greatly dependent on the accuracy of the equilibrium phase diagram and the artistic skill of the user, a quick estimate of 60 to 70% of the lowest represented equilibrium eutectic temperature of the pairwise diagrams is usually sufficient. Above this temperature, reaction can be expected between noncompatible phases. If atomic mobility within and between phases is also sufficient (generally $T > 0.7T_m$) reaction and phase growth may be detectable.

A liquid is not stable at these low temperatures—it is the possible formation of a liquidlike or highly disorganized interface state that may serve as a precursor for phase nucleation, the lower energy condition at the metastable reaction temperature when compared to the continued coexistence of the incompatible phases. The available driving forces may still be insufficient for complex phase nucleation and growth until all species are mobile at much higher temperatures.

Calcination

Calcination is a brief heat treatment of a blended, consolidated, intimate powder mixture with dopants at a temperature of about $0.5T_m$ of an intended product phase. The heat treatment is to promote widespread formation of nucleation sites, often through localized reaction with dopants, for strain-free grain growth during subsequent sintering. Similar effects can be obtained through direct incorporation of seed crystallites in some cases, but in situ seeding through calcination yields a more intimate mixture. Calcination can also simply promote diffusion of dopant species in the near-surface region of a particle which can affect the energetics of initial and intermediate stage sintering. Calcination heat treatment also serves to remove volatile surface impurities and solvent residues from the chemically derived powders which could interfere with the action of subsequently added dopants or with initial stage sintering.

For conventional oxide raw materials, calcination is a prereaction step in a heterogeneous system, and it is particularly effective for the promotion of phase nucleation if the raw materials have been subject to mechanical grinding, usually with a solvent, prior to heat treatment. Surface reaction rates are enhanced by surface-adsorbed solvent in mechanically damaged regions, promoting compound nucleation.

For calcination to have an effect in the control of densification and grain growth during sintering, the nuclei must be well distributed. A calcined powder is normally lightly ground and well blended prior to final fabrication. This step is important as calcination reactions may not occur uniformly in the mass of powder and so the products of reaction must be redistributed evenly prior to sintering or high-temperature reaction. The final aggregate for sintering may well involve additional sintering aids and binders.

The processes of grinding, blending, and calcination can be critical in achieving the goal of abundant and well distributed nuclei in a powder compact. While calcination heat treatment is more specific to control of high-temperature grain growth through the control of lower temperature "prereaction" or nucleation events, appropriate grinding after calcination can introduce additional mechanical flaws which will have a tendency to drive local recrystallization at higher temperatures ($0.7T_m$ or higher). Blending results in a higher degree of homogeneity to ensure an even distribution of prereacted and reactive sites throughout the system volume during the next stage of processing.

Calcination heat treatment of chemically prepared, sol–gel or coprecipitated raw material also serves to improve compaction characteristics and sintering uniformity. A sol–gel precipitate is seldom uniform in either density or chemistry, with a surface structure that may be strongly influenced by the solvent used in formation and the means and completeness of solvent removal. These fine powders will probably be "amorphous to X-rays." However, the bulk or interior of the particles may be fully crystalline, albeit highly strained and defective. During calcination, and after solvent removal, the surface structure will further equilibrate, reducing excess surface activity which could limit sintering and densification.

Also, through Ostwald ripening and some local sintering of sol–gel agglomerates, the finest particle size fractions are eliminated and crystallite nuclei are formed. Ultrafine particles are difficult to consolidate as center-to-center approach is largely controlled through surface charge effects: repulsion will limit compact "green" density. A large, well distributed population of crystallite nuclei and improved particle packing ensures that a more homogeneous, higher density fired microstructure will be obtained on sintering, even without additives.[3]

For intentionally doped systems or those containing sintering aids added prior to calcination, it is reasonable to concentrate on surface composition when choos-

[3]For a quenched, amorphous or "glassy," starting material, well distributed nuclei can result in a very desirable fine-grained microstructure, as found in glass ceramics.

ing the initial system for analysis of calcination events and to examine the phase diagrams of the additive–oxide system(s) to determine whether a potential compound could nucleate during calcination. A reasonable calcination temperature would lie between $0.5T_m$ of that intended or suspected product phase and 0.6 to 0.8 of the lowest eutectic temperature characteristic of the surface system. For mixed oxide powders milled with water or another solvent but without intentional dopants, the pairwise binary diagrams are most suitable for estimation of calcination temperatures.

A potential compound could indicate one of several events could then be expected during calcination which could affect sintering. A readily formed stable or metastable compound of the interface system could serve as a heterogeneous site to nucleate grain growth during sintering. A metastable phase in terms of the broader system composition would be transient and probably undetected in the final sintered microstructure.

Alternately, the nuclei of a stable compound could form which could serve as homogeneous or heterogenous nucleation sites. As a nucleation site, the nature of this phase might alter subsequent grain growth or growth habit. In a nominally single-phase sintering system, the nucleation of an insoluble, stable grain boundary precipitate phase during calcination would help to prevent exaggerated grain growth.

Selection of calcination time and temperature are a matter of some empiricism. A good purity raw material mined at site A will have a different background chemistry than a similar material mined at B, which will be reflected even when each has been purified to the same nominal state. The two materials are likely to have different forming, calcination, reaction, and sintering characteristics. A similar purity "chemically prepared" raw material is likely to differ greatly from either of these in its surface character and sintering behavior. Because of the transient nature of calcination reaction products, the exact sequence and effects on subsequent sintering behavior can only be inferred "after-the-fact" from microstructural and phase development.

Sintering

The majority of ceramic products are sintered, densified, polycrystalline materials. Sintering is defined as the reduction of porosity. Reaction and growth processes may accompany, promote, or hinder the sintering process.

The sintering process in the absence of reaction is the approach to a lower energy state via a reduction of surface energy ($\gamma\, dA$) through densification (reducing solid/vapor interface area). Grain growth (reducing solid/solid interface area) may accompany sintering but is not in itself sintering nor is it required for densification. Excessive grain growth can actually limit densification through the entrapment of porosity.

It is desirable to have a small equiaxial particle size to promote good particle packing and multiple interparticle contacts. The particle surfaces should be unfaceted, as curved surfaces represent in themselves a driving force for atomic motion. The chemical potential of atoms at an interface or boundary is a function of its curvature, with $\mu_{convex} < \mu_{concave}$. Atoms have a thermodynamic tendency to diffuse to the lower energy convex side, with the net boundary or interface motion being toward the center of curvature of the developing interface. A net flux from within the particle toward the contact or "neck" region will result in overall shrinkage and reduction of relative pore volume.

Diffusive transport from the surface region to the neck region (via vapor, surface, or bulk pathways) will result in particle shape change with little densification. Only bulk diffusion and grain boundary diffusion will allow complete densification.

Sintering of Mixtures

A mixture of materials behaves differently than a pure or homogeneous phase in sintering, even in the absence of reaction. For insoluble mixtures, but where mixed particle contact is favored from surface energy considerations ($2\gamma_{AB} - (\gamma_A + \gamma_B) < 0$), a parabolic combination of sintering or shrinkage rates of the pure materials may be observed

$$Y_{mixture} = Y_A C_A^2 + Y_B C_B^2 + 2Y_{AB} C_A C_B$$

where $Y_{A,B}$s are the shrinkage rates observed from the pure materials and $C_{A,B}$s are the concentrations of each component in the mixture. This equation was derived from the behavior of mixtures of bulk materials, but in the initial stage of shrinkage and neck formation, the surface concentrations determine behavior and similar effects will be seen from chemical heterogeneity.

Reactive Sintering

In sintering with reaction, local reaction and phase equilibration is promoted to form a strong and useful product. Useful composite properties can be developed through partial or surface reaction during sintering. Reactive sintering is used, for example, to create boundary layer devices for electronic applications or "reaction bonded" structures for refractory and abrasive application.

The three stages of solid-state sintering may be generalized to sintering with reaction, which is of concern here. The initial stage ($\approx$ 0 to 10% shrinkage) is characterized by the reduction of chemical potential gradients at particle–particle contacts, which may result in nucleation of stable or metastable reaction products. Continued shrinkage into the intermediate stage requires bulk and surface transport to promote coherent and steady growth of the reaction products and steady elimination of vapor–solid interface area. Excessive or exaggerated early grain growth may lead to entrapment of porosity which is difficult or impossible to eliminate in the final stages of sintering which has characteristically only closed and isolated pores (≤ 15% porosity).

For sintering with reaction, the presence of a liquid phase (stable or metastable) promotes initial and intermediate stage shrinkage behaviors, first by allowing limited particle rearrangement through viscous flow and then through promotion of product phase growth through dissolution–reprecipitation and enhanced transport through the liquid. Entrapped gases may also be more readily eliminated as a result of these processes, minimizing or eliminating the enclosed pores characteristic of the final stage of solid-state sintering.

Ceramic powder compacts for sintering are designed, both chemically and microstructurally, to optimize conditions that promote the nucleation and growth of strain-free grains of the desired product phase and also mass transport leading to shrinkage. Ideally, nucleation of reaction leading to grain growth is statistically uniform, controlled by calcination and subsequent blending and careful forming of the green aggregate. Adequate mass transport at high temperature will then ensure that all nucleated grains grow at the same rate with the final grain size being determined by nucleation rather than growth rate.

Absolute certainty that reactive sintering proceeds via purely solid-state reaction is impossible, however, the conditions for solid-state reaction in a multiphase compact may be determined using the phase diagrams appropriate to the solid-state interparticle contacts.

The reactive sintering process is governed by the kinetics of reaction, by mass transport kinetics and surface energy minimization as well as the energetic requirements of phase equilibrium. The behavior of the ceramic during sintering is not fully determined through equilibrium considerations as described in the system phase diagram. Normal sintering practice does not allow for all reactions to go to completion, indeed the sintering schedule (time at temperature) may be designed to select and promote desirable reaction processes that may not represent global equilibrium of the system.

Final stage densification is controlled by diffusion rates of gaseous species from the pore toward the free surface and can be very slow. Pores that are associated with grain boundaries will usually be eliminated most rapidly, as diffusion rates tend to be higher along grain boundaries and within other highly defective regions. Trapped pores cannot generally be eliminated in reasonable sintering times. A natural limit to practical densification, the "endpoint density," is determined by the diffusion rates of such species as N_2, CO, and CO_2, which are very slow. Higher densities can often be observed for the same material when sintered in vacuum or H_2 owing to the elimination of these species, outside of any thermodynamic effect of p_{O_2} on defect related diffusion rates.

A small pore population will always be found in a sintered ceramic. The location and nature of the remnant pores will determine whether their presence will be detrimental to performance. Small pores located only in grain are usually thought to be the least detrimental to mechanical or electrical properties. Any pore that intersects a free surface after machining will be a potential "Griffith" flaw in the determination of fracture strength.

Sintering with a Liquid Phase

Equilibrium liquid formation is not reasonable to expect in a sintering system. An equilibrium liquid would be that predicted from the overall system composition and is characteristic of a fine-scale, statistically homogeneous mixture of the components. The realities of particle size distributions and particle packing statistics ensure that these homogeneous conditions will not be obtained unless fritted (premelted) or prereacted reactant materials are used.

For typical ceramic aggregates containing both coarse and fine particle sizes, uniform distribution of components will be almost impossible. The sintering system will be composed of two or more reacting subsystems. If, by design, one fine-particle or fritted subsystem is low melting, a liquid phase develops that promotes further reaction of the coarser fraction and densification by particle rearrangement. Initial reaction with liquid formation is an essential process in traditional ceramics, such as stoneware or porcelains, which are densified through viscous flow of a significant liquid phase. In that case, the liquid phase is continuous and control of its temperature of formation and viscosity is required to avoid plastic deformation or slump. The initial fluid viscosity is determined by the locally low-melting composition, but will change as dissolution and reaction of more refractory phases proceeds.

In the case of a traditional ceramic, the liquid persists as a glassy phase in the room temperature microstructure. In many cases, however, the liquid phase is transient. A transient liquid disappears either almost immediately as it wets and reacts with the other phases or is more gradually consumed through reaction process. A microstructural characteristic of reaction involving a transient liquid phase is a final grain size which is significantly finer than the starting materials. The relative fineness is due to the nature of the dissolution/precipitation reaction with well distributed nuclei for grain growth. Transient liquids should constitute from 1 to 10% of the system for effective surface coating without excess residual grain boundary liquid phase.

Transient liquid-type behavior for surface reaction and distributed nucleation may be promoted in nominally equiax particle mixtures through the use of precipitated or sol–gel coatings, even having the same "oxide" chemistry as the bulk. The reaction behavior of a coating cannot be simply described with reference to an oxide phase diagram, owing both to the small dimension and the water content of the coating "system." Water (or dissolved H^+ and OH^-) are effective fluxes, promoting low-temperature reaction, which may result in a transient liquid or simply rapid crystallization.

The distinction between a transient surface liquid phase reaction and metastable solid-state reaction is impossible owing to the uncertainties of description of a nanoscale process. However, the conditions of reaction will be consistent with phase equilibria-based predictions for the local system's reacting composition.

Consideration of the Vapor Phase

In an aggregate system for ceramic processing by sintering, particularly if the binding phase is transient or not completely filling of all void space, reaction and phase growth will involve some degree of involvement with the vapor phase. The availability of polar molecules or ions in the vapor may be very important in surface equilibration, restructuring, or defect formation. In reactions where the equilibrium phase is formed at or near the surface, interaction with ambient species may provide a source of electrons or ions to balance local charge gradients.

Solid-state reactions may be observed to proceed differently in response to the atmosphere, being particularly affected by humidity, CO_2, and air "pollutants" such as SO_x. Volatile species from the furnace materials or surroundings may be present at higher temperatures. A volatile species present at pressures higher than about 10^{-3} torr (1 Pa or 10 ppm of normal atmospheric pressure) should probably be considered as a component of the system in view of its probable interactions with the surfaces of fine powders. (Some atmospheric species may be active at much lower pressures.)

Contaminants that will be active in the vapor may also arise from the residues of prereactive or forming steps, particularly metallic residues from cutting tools, lubricants, and binders. These effects are further discussed below.

Sintering Additives

Additives are characterized as dopants, sintering aids, or as binders depending on their intended function. A sintering system will always (and usually intentionally) include a binder and impurities that serve as dopants. Some additives have multiple functions.

Binders include solvents and other, generally organic, transient phases that are used to promote stable rheological or plastic behavior. Binders include metal hydroxides, alkoxides, and hydrated organometallic salts as well as complex hydrocarbons and waxes. In practice, binders make up from 1 to 5 to over 30 vol% of the aggregate system. The latter high percentages are typical for the wax binders used in the injection molding of advanced ceramics. For the highest percentages, the carbon decomposition product is a highly reactive major component of the system. The decomposition products of any of the typical organic binder fractions may also function as dopants or sintering aids.

Surface and fluid water is an almost universal binder, with surface hydroxides serving as practical sintering aids for nucleation in many notional "undoped" pressed powder compacts. Sintering behaviors can reflect the chemistry of water and alcohols adsorbed at numerous process stages.

Sintering aids are nonsoluble reactive impurities or impurities that form phases with the initial oxides which are insoluble in the product phase or phases. Their function is to provide distributed heterogeneous nucleation sites for the desired product phase, limiting grain size through grain impingement, or to limit exaggerated grain growth of that or another phase (otherwise nucleated) by precipitate "pinning" of grain boundaries. Sintering aids are usually added at the 0.5 to 5% range, a similar percentage as dopants. The sintering aid is most successfully added with a solvent or binder system which provides for its uniform distribution on the particle surfaces. Sintering aids may also serve as dopants at high temperature.

Dopants, which may be intentional or nonintentional (impurity) additions, are generally soluble in one or more of the major solid-state phases at the sintering temperature. The dopant promotes defect formation which enhances mass transport. Some dopants are exsolved on cooling from the sintering temperature, or have a tendency to segregate to grain boundary regions (due to lattice strain) with a resultant effect on grain boundary energy, grain shape, or grain boundary phase formation.

It is probably lucky for the advanced ceramics industry that a pure material is thermodynamically unstable relative to a slightly impure and defective phase. A pure single oxide or congruently melting stoichiometric compound would not be a good candidate for sintering. Few (if any) polycrystalline ceramics are truly one-component or simple single-phase multicomponent systems.

Naturally occurring impurities have a statistical likelihood to be distributed throughout the oxide structure, proportionate to their relative electrostatic and strain interactions with the host lattice. Small local energy increases owing to defect formation are more than compensated by increased configurational entropy, so impurity solubility is thermodynamically favored at the ppm level.

A system that is predominantly single-oxide will contain impurities at the level of 100 ppb (10^{-5} mol%) or more, generally 100 ppm (10^{-2} mol%). The low limit is really the effective limit of detection of the most precise analytical techniques; for any single element a practical limit may be somewhat higher or lower depending on the primary oxide and other impurities. Approximately 40 elements have average concentrations of 10 ppm or more in the earth's crust and might be expected to be found somewhat randomly at the ppb level in any material. Other elements, such as tungsten or mercury, are much rarer overall but are commonly used in research and industry. It is thus reasonable to assume that a "pure" material contains at least 10 to 100 ppm total impurities, of perhaps 40 elements, all present but below specific detection limits. For different geographical raw material sources and for different reagent preparation methods, the impurity chemistry will be somewhat unique, affecting defect equilibrium and surface activity.[4]

Dopants, or intentional impurities, are added in the 0.05 to 5 wt% of the batch

[4]These same chemical differences will also result in differing rheological characteristics or solvent interactions, which are not a topic of this work, but can also influence sintering behavior in the compact through effects of particle packing and dopant distribution.

total, or ≥500 ppm. This much exceeds background impurity levels but should be within the solid solution limit as indicated by the relevant phase diagram. Where high levels are used, the dopant is a minor component of the system, particularly if sintering with a liquid phase or a distributed second phase (crystallizing from the liquid on cooling) is desired.

Particle size is often determined as "subsieve size" or by using surface area determination by gas adsorption. An "equivalent spherical particle size" is reported, which is approximately calculated by the relationship:

$$\text{particle size } (\mu m) \approx 6 \div [(\text{s.a., } m^2/g)\cdot(\text{density, } g/cc)]$$

for common materials, a few approximate values are:

	0.5 m²/g	*5 m²/g*	*100 m²/g*
SiO_2(2.2 g/cc)	6 μm	0.5 μm	0.03 μm
Al_2O_3 (4 g/cc)	3 μm	0.3 μm	0.02 μm
ZrO_2 (5.9 g/cc)	2 μm	0.2 μm	0.01 μm

The effects of particle size are enormous on the distribution and effectiveness of dopants (see Box). A uniform powder (A) with 5 m^2/g will have about 16% of monolayer surface coverage at 0.05 wt% dopant (B). A powder having 100 m^2/g surface area will have about 1% of a monolayer coverage. For the 5 m^2/g case, about 16% of interparticle contacts are heterogeneous (A–B–A), 3% homogeneous-dopant (A–B–B–A), and 80% homogeneous-base oxide (A–A). In the latter case only about 1% are heterogeneous or homogeneous-dopant.

In contrast, a coarse powder ("sub 325 mesh" or ~ 0.5 m^2/g) will have complete coverage of all particle surfaces at the same doping level, and all particle contacts except within agglomerates will be homogeneous-dopant.

In the first case, a significant fraction (about 20%) of the system will behave locally as if it had a system composition of 50% additive, whereas in the second case only a small widely distributed fraction of the material will behave in this manner. The relatively heavy coverage in the last case of a coarse powder ensures that all surfaces, except those within agglomerates, are 50% additive or more.

Multiple process cycles of calcination, grinding, and reblending may be necessary to ensure that the desired solid solution, prenucleation, or precipitate formation reaction involving additives will take place and have uniform product distribution.

8.3 Consideration of Engineering Design

A ceramic product is designed for the application of its dielectric, magnetic, optical, structural, and refractory properties which result from the properties of phases and their microstructural arrangement. The engineering properties are characteristic of a stable nonequilibrium state of the system.

The sintered ceramic component has been at least partially equilibrated at the sintering or subsequent annealing temperature. Its characteristic room temperature state will be that of a kinetically stranded system. Similarly, a crystalline or glassy product of fusion is a kinetically stranded system.

Provided that the material was first equilibrated under known conditions, a kinetically stranded material and its temperature range of relative stability is defined with reference to the equilibrium phase diagram. Kinetically stranded materials will be characterized by chemical potential gradients and structural defects which allow significant energy reduction on a local scale. The most reasonable of these local structures are those that are closely related to atomic structures typical of the equilibrium system, again determined with reference to the phase diagram.

The equilibrium phases are characterized by both long- and short-range ordering which represent a minimum in free energy. Under conditions in which the bulk material phases are nonequilibrium or stranded, local atomic arrangements such as at surfaces and interfaces may show "near" equilibration. Owing to segregation, additive distribution, and other processing details, interface compositions may be much different than overall or average compositions. At surfaces and interfaces where volume constraints are minimal and defects, such as from impurities, are plentiful, local structure and properties may be very similar to the macroscopic equilibrium phases as given by the phase diagram for the local chemical system. Continued irreversible reaction or progress toward equilibrium is favored under any conditions where process heat or energy gained from internal loss mechanisms is sufficient for atomic motion.

Thermal Interaction and "Aging"

The as-fabricated ceramic is not at equilibrium, either internally or with its environment, but is stable against damaging alteration in application at or near ambient temperatures. Ceramic products designed for the nonindustrial consumer are used without question of stability, generally considered inert and unchanging at $T_h < 0.2$ to 0.3, at least 200 to 300°C (600°F). In earlier discussion, mechanically bonded heterogeneous composite systems were also described as generally stable under these conditions.

For higher temperature system applications, the phase diagram is particularly valuable in the anticipation of events that will occur after processing. The present discussion will assume that the as-fabricated ceramic is mechanically stable, unflawed, and crack-free. It is reasonable to assume that the mechanical, thermal, dielectric, and magnetic properties are structurally isotropic owing to a nominally equiax and randomly oriented grain structure.

The ceramic product is a component of an heterogeneous operating system that will be used in an environment and is expected to be stable for a significant service lifetime. During that lifetime, the ceramic component will see interaction with electronic and magnetic fields as well as thermal, chemical, electrochemical, and mechanical stress gradients. The first utility of the phase diagram is for failure avoidance, in determination of the "safe" temperature range for continuous long-term operation.

High-temperature ceramic interaction with the operating system and its environment has several general outcomes.

First, neither the ceramic component nor the engineering system have significant alteration of properties, or the properties of the system may be improved by the interaction. This would be called the result of "breaking in" the device. For response to an electromagnetic field or electrical potential, "break-in" corresponds to the initial reduction of junction potentials that exist between the dissimilar materials in the as-fabricated device. Better response times in cyclical measurement systems, such as in oxide sensors and photochromic glasses, develop during break-in owing to defect and structural relaxations or local cluster formation. These behaviors are not generally governed by phase equilibrium considerations.

Second, the ceramic may be progressively altered without serious effect on system performance. This would be the case for thermal insulations that are not exposed to corrosive environments, where the required property is dimensional stability and load bearing hot strength at moderate temperatures, $T_h \geq 0.3$ to 0.5. Although local structural equilibration will occur in and among the refractory phases at the higher end of this temperature range, no extended deformation via creep or plastic flow would be expected to occur.

Extended operation of electrolytic devices based on refractory zirconia and alumina phases at $T_h \approx 0.5$, will result in "aging" or reduction in peak response or output, as mobile defects or excess solute (for the operation temperature) cluster or are precipitated as a second phase. Aging behaviors that are related to exsolution of solute or a second phase can be "recoverable," in whole or in part, by reannealing the ceramic component at a temperature in the homogeneous solid solution range, as determined using the phase diagram.

"Aging" can also be due to progressive interaction with heterogeneous contact materials at these temperatures. Reference to the phase diagram for the system most characteristic of the heterojunction can aid in understanding this behavior and avoiding more serious effects where possible. For instance, while a contact metal may go into metastable solution in the ceramic, degrading the dielectric properties, reference to the phase diagram will indicate whether a higher temperature excursion could act to promote precipitation of the impurity as a second phase or in an extended defect cluster, "cleaning" the structure. (Some sintering aids are also active in the promotion of grain boundary and defect structures that have a high affinity for metallic and environmental impurities, reducing their deleterious effects.) The indication of stable solution on the phase diagram would indicate that high temperatures were to be avoided.

At these moderate temperatures, $T_h \approx 0.5$, limited chemical interaction between heterogeneous phases may occur which results in an altered local phase equilibria while benefitting short-term stability of an operational system. Formation of solid–solid–vapor corrosion products can effectively seal a refractory block wall, although the continued interaction, or interaction at a higher temperature, will destroy the ceramic. Initial corrosive interaction is often well described using the phase diagram of the "fluxed" system or systems that characterize the local condensing vapor–solid surface phases at the surface temperature. A coherent product will also tend to passivate the surface and protect against further interaction.

(Oxide ceramics are also subject to corrosive damage due to long-term hydration reaction at ambient temperatures, as evidenced by efflorescence, pitting, and spalling of structural ceramic surfaces. Normal "ambient" temperature and relative humidity are sufficient for metastable liquid formation in the water–salt systems responsible for the progressive reaction, although major damage is usually brought about through secondary freeze–thaw mechanical disruption of the near surface.)

Nonthermal Interactions

The available energy includes not only heat, as quantified by the homologous temperature, but also energy due to interaction of the ceramic and its system with energetic fields.

Oxides are dielectric materials with physical, electronic, and vibrational states that respond to applied electromagnetic fields and electrical as well as chemical potentials. In alternating fields, each ceramic system and its contact materials form a local impedance network with a characteristic loss response. The dielectric loss represents an available energy-per-cycle for local transitions, defect transport, and interfacial reaction which is in addition to the thermal energy as measured by temperature.

Similarly, magnetic switching losses or eddy current losses (due to free electrons) contribute to localized heating effects. Soft ferrite microstructures, which are designed for rapid switching, must be designed to minimize features such as precipitates or in-grain porosity that tend to pin magnetic domain walls.

Vibrational and acoustic interactions lead to internal friction losses and local heating and cyclical compressive stresses. Direct surface frictional or tribological interaction can lead to significant chemical reaction, indicating very high local temperatures and pressures.

These cyclical loss processes may increase the effective "temperature" significantly, as measured by the vibrational and electronic states of the individual atoms, allowing reaction processes to occur at a lower temperature than expected from the phase diagram. Microscopic and progressive changes can be brought about owing to localized energy lost per cycle, even near room temperature.

Static and varying DC field interactions may be responsible for larger scale microstructural alteration at moderate temperature. The imposed potential, which will vary from point to point in the microstructure, superimposes upon the inherent chemical potential gradients characteristic of the material, influencing the "drift" current of ions and charged defects and can lead to gross microstructural changes such as preferential or oriented grain growth. Microstructural alteration, grain growth, and diffusion can all lead to the eventual formation of "short-circuit" pathways through the dielectric.

Any imposed field interacts more strongly with microstructural imperfections and heterogeneities. For electromagnetic fields this may result in localized Joule heating which can lead to runaway thermal effects and breakdown.

Interactions to Failure

At high temperatures or under high field conditions at moderate temperatures, the ceramic can be altered with deleterious effects on the system or the process, both through contamination (as from a melt-containment interaction) and through physical failure and system collapse. Extensive interaction is a trend toward long-term equilibration of the ceramic with its environment and is typical at $T_h > 0.8$ (relative to melting point or ceramic system eutectics). Interaction is not strictly chemical. In load-bearing installations subject to high electromagnetic fields, lossy behaviors due to metallic carriers in the nominally insulating ceramic can lead to localized heating, recrystallization and crack initiation.

At these high temperatures, the ceramic's properties become functionally dependent on environmental interactions. After a long service period of equilibration with an active environment, the phase equilibria that described the original refractory are unlikely to describe its chemically and physically altered condition. Local temperature, chemical potential gradients, and mass flow rates may vary widely within a single installation, as well as electromagnetic fields and mechanical/vibrational stresses. Therefore, the mode and degree of alteration will vary from point to point in an operating system.

Progressive nonlocalized physical alteration leading to failure is typical of glass- or fluid metal-contact refractory alteration. Fluid contact ensures large-scale mass transport, so that the reaction products may be continuously removed and deposited at long distances from the initial reaction site. Interaction in this case includes nonequilibrium reaction or dissolution in low flow rate systems. In high flow rate systems, physical erosion and cavitation may also occur. These installations are also subject to high melt-face temperatures and steep thermal gradients, which can lead to microstructural changes within the ceramic at a distance from the reacting interface.

The maximum service temperature of a ceramic is defined by the environmental, mechanical, and chemical stresses present. Use temperatures are often reported this way in commercial literature, with statements such as "no-load" service temperature or "in noncorrosive environments." Under conditions of "no-load," a small (<1 to 5 vol%) liquid phase may be tolerable and the maximum use temperature can be defined directly from an appropriate phase diagram for the base system. Each percent of added component generally will lower the temperature by 20 to 50°C, from the general trends predicted by the "freezing point depression."

MgO-based refractories illustrate this point. Fused-cast, 97% MgO block has a maximum use temperature of 2700°C, 100°C lower than the melting point of pure MgO. Lower purity "dead burned magnesia," 83 to 93% MgO, is limited to use below 2200°C. These use temperatures reflect a freezing point depression of 35°C/mol of impurity in general. An added margin for safety is added in the latter case as the major impurity is Fe_2O_3 at room temperature in air, which may be reduced to FeO at elevated temperature in reducing atmosphere. Reduction or disproportionation of an oxide constituent similarly contributes to the limit on the use temperature of "Mag-Chrome" refractories, which are nominally formed in the high MgO corner of the $MgO–Al_2O_3–Cr_2O_3$ system in the region MgO + Spinel (ss). This type of refractory is said to fail at 1650°C, closely corresponding to the eutectic in the reduced system $Cr–Cr_2O_3$. More general consideration of the lowest eutectic, about 1900°C between MgO and $MgAl_2O_4$, and the 10 to 20% impurity content would result in a lower safe maximum use temperature under load.

Similarly, aluminosilicate refractories such as firebrick (low alumina side of mullite) and high alumina brick (high alumina side of mullite) have differing maximum temperatures for no-load conditions which reflect liquidus behavior, with maximum use at about 1750°C and 1850°C, respectively. However, under load conditions, both are limited by the behavior of the residual glassy phase which flows at about 1350°C, resulting in subsidence.

Thus, excursions to near the minimum melt temperature may be tolerable under relatively stress-free, no-load, conditions. It is possible, particularly where a glassy phase is present, that temperature excursions above T_g into the annealing range may serve to heal small flaws and improve mechanical service characteristics.

Liquid Containment Interactions

Liquid containment at high temperatures is probably the most studied refractory application problem. Even where the liquid–refractory interaction is apparently well understood, such as in the refining of pure metals, concern for the physical stability of the system is paramount to provide for operator safety in case of catastrophic failure. We see the common use of redundant refractory layers in containment structures and incorporation of cooling systems to freeze-off leaks which inevitably may occur.

A typical and demanding containment application using ceramic refractory materials is the glass melter, diagrammed in Figure 8.2. The overall design involves several materials, all of which must be compatible over a wide tempera-

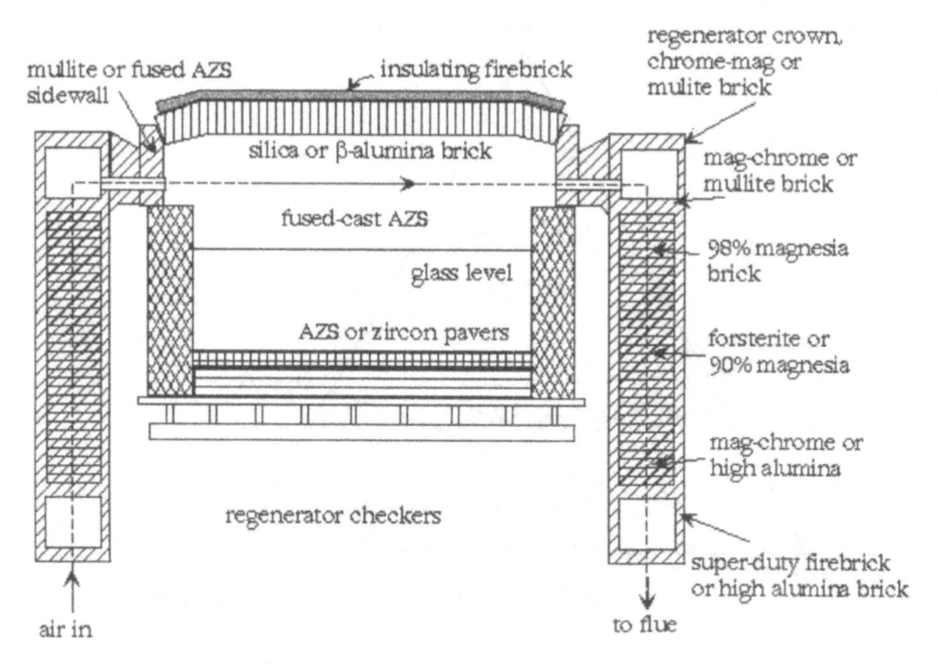

Figure 8.2 Simplified cross-section of glass tank refractory installation in the production environment. Structural integrity, thermal insulation, and resistance to chemical and ablative attack are required for adequate service.

ture range while under mechanical constraint and subject to chemical interaction and frictional or ablative attack.

The ideal refractory materials for melt contact in an open system are not necessarily chosen by a single criterion such as melting point. As an example, chrome oxide or mag-chrome refractories are extremely durable in the glass tank environment but see limited application because of strong coloration of the glass from trace amounts of dissolved chrome as well as the toxic waste hazard they present on disposal of the refractory itself after its service lifetime.

Commonly used AZS (alumina–zirconia–silica) refractories are a compromise in performance characteristics. Their composition range is shown in Figure 8.3, indicating their ready fabrication from relatively inexpensive raw materials. Alumina and silica are soluble in the glass, however zirconia tends to form zircon stones which are stable. Zircon is more dense than the glass melt, however, so limited carryover is seen to the finished flat glass sheet. Any deleterious stones, bubbles, or other flaws are routinely located, eliminated from production, and recycled with cullet (crushed glass scrap). The economies of scale in using the AZS refractory for long tank life more than account for any additional scrap generation.

High alumina refractories, although less durable, also see application as dis-

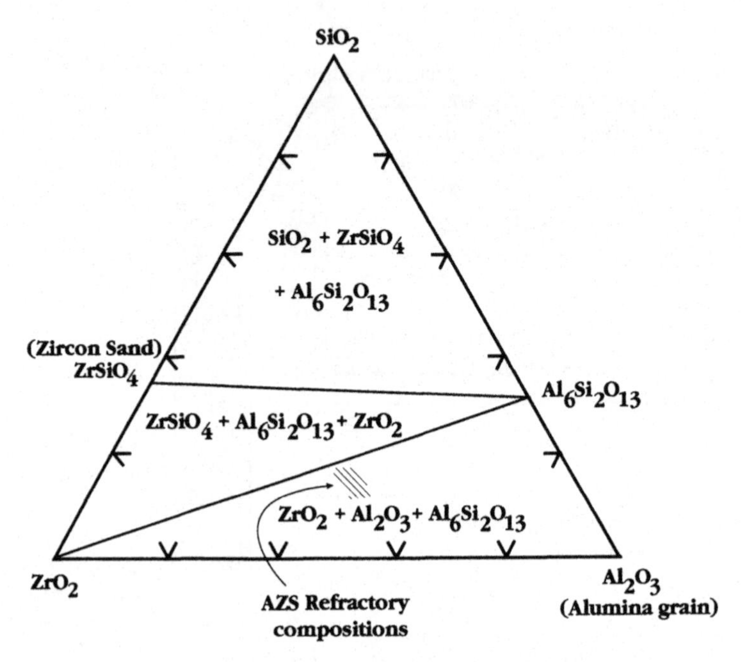

Figure 8.3 Schematic alumina–zirconia–silica phase diagram. AZS composition range is shaded, nominal raw materials circled. The indicated equilibrium mullite phase does not crystallize and forms the glassy phase of the refractory.

solved alumina does not significantly alter the properties of the glass, and are particularly suited for optical glass production. Similarly, silica and β-alumina bricks are applied in crown (roof) construction, as any liquid formed from vapor attack that drops back into the glass will be quickly incorporated.

Refractory applications in metals refining are sketched in Figure 8.4. In metals production it is common to use "tar bonded" oxides or carbide/oxide composites whose two phase persistence fixes the oxygen potential of the molten metal contained. Some oxide refractories are designed to be consumable, producing slags with impurities in the metallic ore which are then separable by floatation from the relatively denser metal. Viscous slags also serve to protect the refractory surface from further interaction in metals contact as well as glass contact applications. Any dissolution or slagging reaction requires continuous monitoring as failure tends to be catastrophic.

Limited interactions in containment of molten systems can be beneficial to the operation of the system. For instance, initial reaction may effectively seal seams and joints in between refractory blocks or tiles, preventing leakage and corrosion of the outer containment. A reaction product can also serve to passivate the surface or buffer further reaction.

The reaction processes observed on a large scale in refractory containment

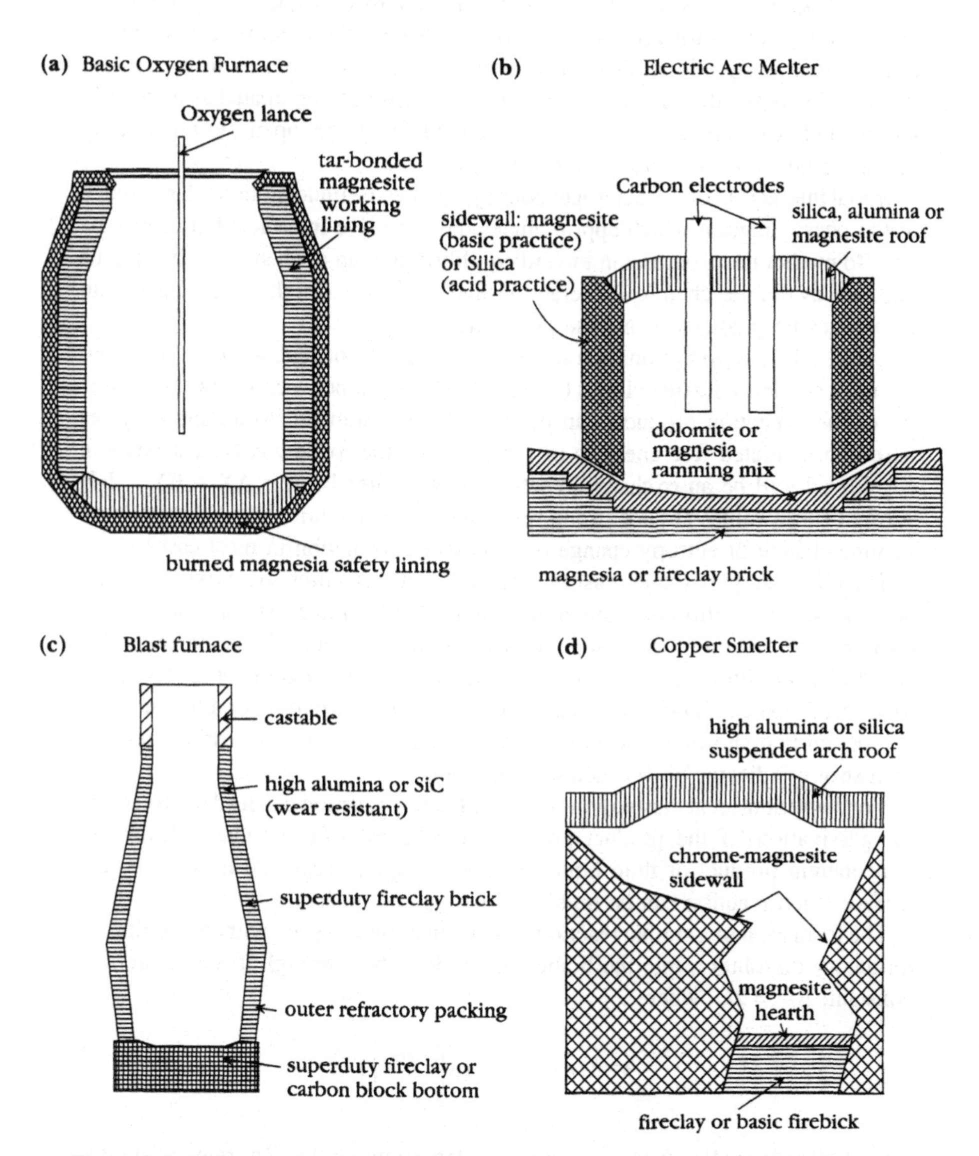

Figure 8.4 Refractory installations for metal refining. (a) Basic oxygen furnace (BOF); (b) electric arc furnace; (c) blast furnace; (d) copper smelter. (Drawing are schematic and not to scale.)

systems are similar in nature to those observed in liquid phase and reactive sintering on a smaller scale. On both macro and micro scales, the reaction process and resulting microstructures will be consistent with the requirements of phase equilibrium. A significant difference between these processes is in the system description. While the calcination or sintering system can be treated as nominally closed and isothermal, the refractory containment is an open system usually operating under steep thermal gradient conditions.

Initial interaction of the heterogeneous system can be understood with reference to the phase diagram which approximates the corrosive interface, but cannot be used to predict the progression toward equilibrium in an open system. If liquidus data are available, chemical interaction and reaction at high temperatures can sometimes be predicted using the phase diagram.

Often, a useful approximation on relative tendency of reaction between more complex condensed oxide phases (minerals or slags) or between various refractories or their reaction products can be made using estimated formation energies based on tabulated free energies of formation of the oxides. A typical reaction considered will be an exchange reaction of the general form: AX + BY = BX + AY. (These values are not strong functions of temperature, nor is there a large volume change or entropy change on compound formation in most cases.)

The most complex oxide must be formable from its simple base oxide constituents at some equilibrium temperature, allowing a simple estimate of the free energy of formation of the compound from tabulated data. The estimated value can be used, with caution, to evaluate each proposed reaction for its tendency to proceed (ΔG_{rxn}). A possible reaction leading to the formation of a low melting compound that shows a tendency to proceed ($\Delta G_{rxn} < 0$) by this estimation will be a strong indicator of destructive interaction.

Condensation-local dissolution–recrystallization reaction can lead to protection (or passivation) if the product phase is a coherent solid or glassy layer. As noncoherent product or fluid liquid product is rapidly removed from a vertical surface, such reaction can be highly damaging and erosive.

The formation of a simple reaction product in a static system without significant refractory dissolution into the liquid can be described through its parabolic rate constant, $\Delta x^2 = 2\bar{k}t$, where

$$\bar{k} = D_{\text{slow}}\left(1 - \exp\left(\frac{n\Delta G_f}{k_b T}\right)\right)$$

The product layer thickness, x, increases proportional to $t^{1/2}$. The rate constant is dependent on both the free energy of formation of the reaction product and the rate-controlling diffusional species, the slowest species in its "slowest path." In the relationship, n is a very small factor, usually varying from 1 to 6, depending on the transport regime in the solid product layer.

A solid product may also form simultaneously with dissolution of the refractory surface. In that case, a limiting thickness may sometimes be obtained that reflects the competing processes of solid-state diffusion and product formation at the refractory–product interface and dissolution at the solid product–liquid interface.

Any positive flow, due to forced convection or gravity, will tend to increase the dissolution rate and will lead to a change from partial passivation behavior toward true corrosion.

Corrosion or reaction in a flowing liquid with velocity u or under laminar (low) flow or gravity driven conditions can be described as

For laminar flow:

$$\frac{\left(\frac{dn}{dt}\right)}{A} \propto u^{1/2}\nu^{1/2}\Delta C$$

For vertical flow under gravity:

$$\frac{\left(\frac{dn}{dt}\right)}{A_{avg}} \propto \left(\frac{g\Delta\rho}{\rho^0 h\nu}\right)$$

where

$$\frac{\left(\frac{dn}{dt}\right)}{A} = \text{mass removal rate per unit area}$$

ν = kinematic viscosity, g = gravitational constant
$\Delta\rho = \rho_{soln} - \rho_o$ and $\Delta C = C_{interface} - C_{\infty}$.

The formalism of corrosion in a flowing liquid, above, incorporated a dissolution rate that is proportional to the relative degree of unsaturation in the liquid. The formation of a saturated, viscous liquid that remains localized at the interface region, as is often observed in high silica slags, serves to limit reaction of the refractory.

Corrosion interaction is progressively further removed from phase equilibrium as the interface liquid velocity increases and dissolution into a large volume or continuous flow eliminates realistic consideration of even a nominally closed system. However, simultaneously within the refractory, progressive phase changes occur that can often be predicted and can always be interpreted in terms of phase equilibrium of the local chemical systems. Prolonged survival in an aggressive environment is limited by not only the nature and degree of the initial chemical and physical interaction but also by the durability of the interaction product against that same aggressive environment.

References

Ceramic Processes and Processing

R. J. Brook, R. W Cahn, and M. B. Bever, editors (1991). *Concise Encyclopedia of Advanced Ceramic Materials.* Pergamon Press, Oxford, New York, and Cambridge.

W. D. Kingery, H. K. Bowen, and D. R. Uhlmann (1976). *Introduction to Ceramics,* 2nd edition. John Wiley & Sons, New York.

J. S. Reed (1988). *Introduction to the Principles of Ceramic Processing.* John Wiley & Sons, New York.

Thermodynamics, Crystal Chemistry, and Physics

P. W. Atkins (1986). *Physical Chemistry, 3rd edition.* Oxford University Press, Oxford (UK).

R. M. Garrels and C. L. Christ (1965). *Solutions, Minerals and Equilibria.* Harper & Row, New York.

C. Kittel (1986). *Introduction to Solid State Physics.* John Wiley & Sons, New York.

H. A. J. Oonk (1981). *Phase Theory: The Thermodynamics of Heterogeneous Equilibria.* Elsevier Science Publishers. New York.

Surendra K. Saxena, editor (1983). *Kinetics and Equilibrium in Mineral Reactions.* Springer-Verlag, New York.

J. C. Slater (1967). *Quantum Theory of Molecules and Solids, Vol. 3: Insulators, Semiconductors and Metals.* McGraw-Hill, New York.

Phase Equilibria and Phase Diagrams

J. M. Sangster and A. D. Pelton (1989), "Critical Coupled Evaluation of Phase Diagrams and Thermodynamic Properties of Binary and Ternary Alkali Salt Systems," pp. viii–xiv

(Introduction II) in *Phase Diagrams for Ceramists, Vol. VI,* edited by L. P. Cook and H. F. McMurdie. The American Ceramic Society, Columbus, OH.

A. Findlay and A. N. Newton (1951) *The Phase Rule and Its Applications*, 9th edition, edited by A. N. Campbell and N. O. Smith. Dover Publications, New York.

R. Haase and H. Schönert (1969). *Solid Liquid Equilibrium.* Topic 13, Vol. 1 of *The International Encyclopedia of Physical Chemistry and Chemical Physics.* Pergamon Press, London.

C-K. Kuo, T-H. Lin, and T. S. Yen (1990). *High Temperature Phase Equilibria and Phase Diagrams.* Pergamon Press, New York (esp. Ch. 7).

E. M. Levin, C. R. Robbins and H. F. McMurdie, editors (1964), "General Discussion of Phase Diagrams," the Introduction to *Phase Diagrams for Ceramists, Volume I.* The American Ceramic Society, Columbus, OH.

J. S. Marsh (1935). *Principles of Phase Diagrams.* Published by McGraw-Hill, New York for the Engineering Foundation (US).

R. E. Newnham (1978), "Phase Diagrams and Crystal chemistry," pp. 1–73 in *Phase Diagrams: Materials Science and Technology, Volume V, "Crystal Chemistry, Stoichiometry, Spinodal Decomposition, Properties of Inorganic Phases,"* edited by A. M. Alper. Academic Press, New York.

T. P. Seward III (1970),"Metastable Phase Diagams and Their Application to Glass Forming Systems," pp. 295–338 in *Phase Diagrams: Materials Science and Technology, Vol. I, "Theory, Principles and Techniquesof Phase Diagrams,* edited by A. M. Alper. Academic Press, New York.

F. E. W. Wetmore and D. J. LeRoy (1969). *Principles of Phase Equilibria.* Dover Publications, New York. [coverage of 4-component systems]

W. L. Worrell and J. Hladik (1972), "Thermodynamic Equilibrium Diagrams," pp. 747–798 in *Physics of Electrolytes, Vol. 2, Thermodynamics and Electrode Processes in Solid State Electrolytes,* edited by J. Hladik. Academic Press. New York.

Phase Diagrams of Nonmetallic and Ceramic Systems

Phase Diagrams for Ceramists, Vols. I–VIII The American Ceramic Society. Columbus, OH:

I (1964) Edited by E. M. Levin, C. R. Robbins, and H. F. McMurdie.

II (1969) Edited by E. M. Levin, C. R. Robbins, and H. F. McMurdie.

III (1973) Edited by E. M. Levin, and H. F McMurdie.

IV (1981) Edited by R. S. Roth, T. Negas, and L. P. Cook.

V (1983) Edited by R. S. Roth, T. Negas, and L. P. Cook.

VI (1987) Edited by R. S. Roth, J. R. Dennis, and H. F. McMurdie.
[Vols I–VI are nonspecialist, including mostly oxide and metal + oxide systems of interest in ceramic applications]

VII (1989) Edited by L. P. Cook and H. F. McMurdie.
[Halide systems, many calculated diagrams with methods discussed]

VIII (1990) Edited by B. O. Mysen.
[Geological, high pressure, and hydrothermal systems]

Under a new series title, but continuous numbering, *Phase Equilibria Diagrams, Vols. IX–XII*:

IX (1992) "Semiconductors and Chalcogenides," edited by G. B. Stringfellow.

X (1994) "Borides, Carbides and Nitrides," edited by A. E. McHale.

XI (1995) "Oxides," edited by R. S. Roth.

XII (1996) "Oxides," edited by A. E. McHale and R. S. Roth.

Also a part of this series; *Phase Equilibrium Diagrams, Annuals '91, '92 and '93.* edited by A. E. McHale
[The annuals include may mixed systems that did not "fit" the *titled* volumes (IX–XII) as well as a significant number of complex oxide systems]

and *Phase Diagrams for High T_c Superconductors,* Edited by J. D. Whitler and R. S. Roth. (1991).

Defects, Reaction, and Growth

C. R. A. Catlow and W. C. Mackrodt, editors (1982). *Computer Simulation of Solids.* (#166 of the series "Lecture notes in Physics") Springer-Verlag, New York.

M. J. Dignam (1972), "The Kinetics of the Growth of Oxides" pp. 92–286 in *Oxides and Oxide Films, Volume 1,* edited by J. W. Diggle. Marcel Dekker. New York.

V. I. Dybkov (1986a), "Reaction Diffusion in Heterogenous Binary Systems, Part I." *J. Mat. Sci.* **21**:3078–3084.

V. I. Dybkov (1986b), "Reaction Diffusion in Heterogenous Binary Systems, Part II." *J. Mat. Sci.* **21**:3085–3090.

T. J. Gray, D. E. Rase, R. R. West, D. P. Detweiler, W. G. Lawrence, and T. J. Jennings (1957). *The Defect Solid State.* Wiley–Interscience Publishers, New York.

F. A. Kroger (1974), "The Chemistry of Imperfect Crystals," in *Applications of Imperfection in Chemistry; Solid State Reactions and Electrochemistry, Vol. 3.* North Holland, New York.

F. J. J. van Loo (1990), "Multiphase Diffusion in Binary and Ternary Solid State Systems," *Prog. Solid State Chem.* **20**:47–99.

H. Schmalzried (1981). *Solid State Reactions.* Volume 12 of *Monographs in Modern Chemistry.* Verlag Chemie GmBH. Weinheim.

G. H. Tamman (1925). *The States of Aggregation; the Changes in the States of Matter in Their Dependence upon Temperature and Pressure.* Van Nostrand, New York. [translated from the original German by R. F. Mehl.]

W. A. Tiller (1970). "The Use of Phase Diagrams in Solidification," pp. 199–244 in *Phase Diagrams: Materials Science and Technology, Vol. I, "Theory, Principles and Techniques of Phase Diagrams,* edited by A. M. Alper. Academic Press, New York.

Sintering and Microstructure Development (see also general references on ceramic processing)

F. M. d'Heurle (1988). "Nucleation of a New Phase from the Interaction of Two Adjacent Phases." *J. Mat. Res.* **3**:167–195.

D. L. Johnson and I. B. Cutler (1971). "The Use of Phase Diagrams in the Sintering of Ceramics and Metals," pp. 265–291 in *Phase Diagrams: Materials Science and Technology, Vol. II, "The use of Phase diagrams in Metal, Refractory, Ceramic and Cement Technology,"* edited by A. M. Alper. Acadmic Press, New York.

D. Kolar (1991), "Chemical Reaction Controlled Microstructures and Properties of Ferroelectric Ceramics," pp. 3–20 in *NIST Special Publication #804, Chemistry of Electronic Ceramic Materials*, edited by P. K. Davies and R. S. Roth. U.S. Department of Commerce.

Electronic and Magnetic Ceramic Properties and Applications

J. C. Bean et al. (1990). *Epitaxial Heterostructures.* Proceedings of the Symposium, April 16–20, 1990, San Francisco CA. Materials Research Society, Pittsburg.

J. Crangle (1977). *The Magnetic Properties of Solids.* Williams Brothers Ltd., Merseyside (UK).

B. C. Cullity (1972). *Introduction to Magnetic Materials.* Addison-Wesley, Reading MA.

T. Kudo and F. Fueki (1990). *Solid State Ionics.* Kodansha (Tokyo) and VCH Publishers, New York.

A. J. Moulson and J. Herbert (1990). *Electroceramics: Materials, Properties and Applications.* Chapman and Hall, New York.

R. E. Newnham (1986), "Composite Electroceramics." *Ferroelectrics* **68**:1–32.

V. E. Yurkevich and B. N. Rolov (1986), "Thermodynamical Model of Phase Transitions in Ferroceramics." *Ferroelectrics*, **68**:265–274.

Ionic Bonding and Structures

I. D. Brown (1978), "Bond-Valences—a Simple Structure Model for Inorganic Chemistry." *Chem. Soc. Rev.*, **7**:359–376.

I. D. Brown and R. D. Shannon (1973), "Empirical Bond Strenth–Bond Length Curves for Oxides." *Acta Cryst.* **A29**:266–282.

R. D. Shannon and C. T. Prewitt (1969), "Effective Ionic Radii in Oxides and Fluorides," *Acta Cryst.* **B25**:925–926.

R. D. Shannon (1993), "Dielectric Polarizabilities of Ions in Oxides and Fluorides." *J. Appl. Phys.* **73**:348–366.

Containment and Refractory Ceramic Applications

T. Emi (1976), "Slag-Metal Reactions from the Electrochemical Viewpoint," pp. 277–332 in *Electrochemistry,* edited by J.O.M. Bockris. Butterworths, London.

H. M. Kraner (1971), "The Use of Phase Diagrams in the Development and Use of Refractories," pp. 67–115 in *Phase Diagrams: Materials Science and Technology, Vol. II, "The use of Phase diagrams in Metal, Refractory, Ceramic and Cement Technology,"* edited by A. M. Alper. Academic Press, New York.

J. van Muylder (1981), "Thermodynamics of Corrosion," pp. 1–96 and 539–544 in *Comprehensive Treatise on Electrochemistry, Vol. 4,* edited by J. O. Bockris, B. E. Conway, and E. Yeager. Plenum Press, New York.

W. L. Worrell and J. Hladik (1972), "Thermodynamic Equilibrium Diagrams," pp. 747–798 in *Physics of Electrolytes, Vol. 2, Thermodynamics and Electrode Processes in Solid State Electrolytes,* edited by J. Hladik. Academic Press, New York.

Suggested Exercises

Chapter 3

3.1 Write defect incorporation reactions for

CaO in MgO
Na_2O in MgO
P_2O_5 in Al_2O_3
SiO_2 in P_2O_5

3.2 The ionic radii of the species above are collected in the table. Values are given for different anion coordination where appropriate:

	Ionic radius (Å) in anion coordination		
Cation	4-coord	6-coord	8-coord
Na (1+)	0.99	1.12	1.18
Ca(2+)	—	1.00	1.12
Mg(2+)	—	0.72	0.89
P(5+)	0.17	0.38	—
Al(3+)	0.39	0.54	—
Si(4+)	0.26	0.40	—

[Values from R.D. Shannon, (1976). *Acta Cryst.* **A32**:751–67]

Using crystal data or other sources, determine the appropriate substitutional cation coordination. Which substitutional solid solutions would you expect to be stable?

3.3 Write the electroneutrality equations for each of the solid solutions in question 3.1.

Chapter 5

5.1 Define a Phase.

5.2 Define a Component.

5.3 What are the maximum number of phases that exist in a system having three components?

5.4 What is the maximum variance of a two-component system? Of a three-component system? Discuss the meaning of these values.

Chapter 6

6.1(a) Estimate the conditions of the metastable invariant equilibrium $Q_1(s) + Q_3(s) + L_2$.

(b) Estimate the vapor pressure over metastable Q_2 at 100 K.

(c) Estimate the vapor pressure over metastable Q_3 at 100 K.

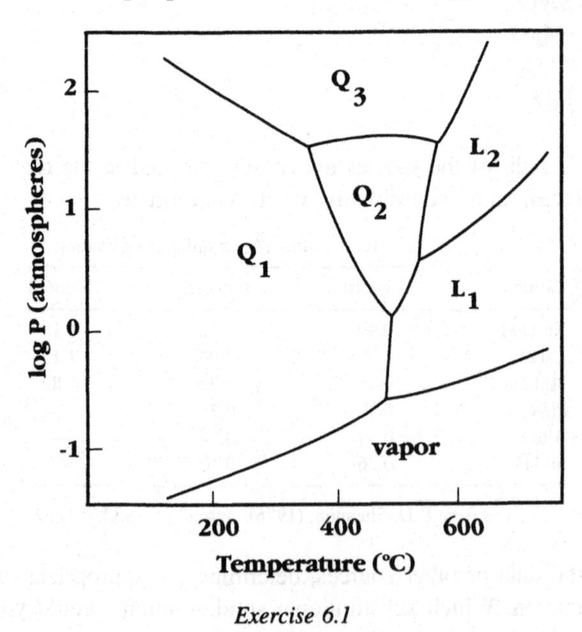

Exercise 6.1

6.2 Draw a single-component diagram consistent with the following information:

There are two solids, S_1 and S_2, and two liquids, L_1 and L_2. there is one vapor phase.

S_1 is denser than L_2 which is denser than S_2.
L_1 is denser than L_2.
S_1 can melt only.
S_2 can sublime and melt.
The reaction $S_2 \rightarrow S_1$ is exothermic.

6.3 Consider systems of the type $MgO \cdot AO_2$ where A = Ti, Mn, Nb, Si, Zr, Hf, Pb. Checking data sources such as diffraction data (JCPDS) and phase diagram compilations (Ceramic Phase Diagrams), look for patterns in phase formation.
Could you suggest a basis on which to "group" similar systems in this situation?

6.4 Consider the oxidation reactions:

$$Ti + O_2 \rightarrow TiO_2 \qquad \Delta G_{rxn} = -212{,}600 \text{ cal/mol}$$
$$W + 3/2O_2 \rightarrow WO_3 \qquad \Delta G_{rxn} = -182{,}620 \text{ cal/mol}$$
$$2Li + 1/2O_2 \rightarrow Li_2O \qquad \Delta G_{rxn} = -134{,}130 \text{ cal/mol}$$

(all forming crystalline solids)

Add these data to the Ellingham diagram of Figure 6.4. Are the lines you have drawn valid at all temperatures of the diagram?

6.5(a) Label all two-phase regions of the diagram for the system ZrO_2–La_2O_3.

(b) Rescale the diagram for the ZrO_2–La_2O_3 system in terms of wt% components.

Molecular weights: ZrO_2 = 123 g/mol $\quad$ La_2O_3 = 326 g/mol

6.6(a) Apply the lever rule to determine the equilibrium phase proportions in the following systems:

At 1400°C, the system 60 mol% A, 40 mol% B.
At 1400°C, the system 20 mol% B, 80 mol% A.

(b) At what temperature does the last liquid crystallize from the composition containing 30 mol% A, 70 mol% B. What is the composition of that liquid?

(c) Indicate the crystallization types of the diagram.

(d) Draw phase analysis diagrams for systems containing (1) 60 mol% A and (2) 27 mol% A.

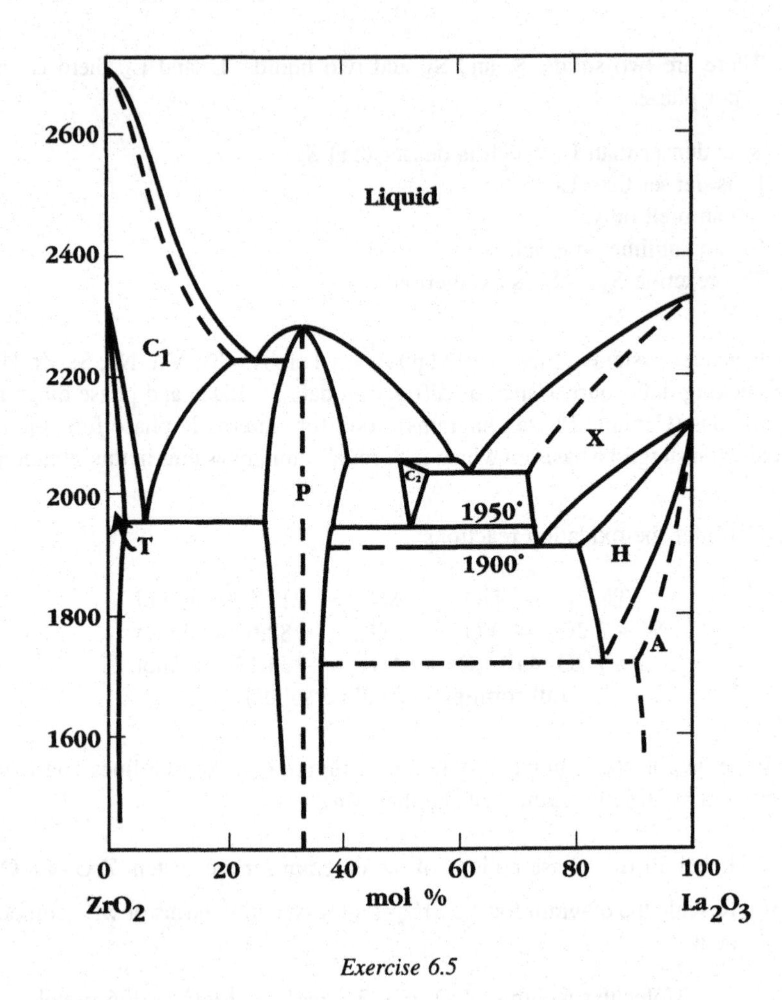

Exercise 6.5

(e) Redraw the diagram to indicate about 5% solid solution of either component in all compounds and endmembers. Be sure to consider the effects of phase transition on the saturation boundaries of the compound AB.

6.7 Calculate a simple eutectic diagram for components X and Y where

	X	Y
T_m	2100°C	1600°C
ΔH_{fusion}	4 kJ/mol	3 kJ/mol

use $R = 8.14$ J/mol-K.

6.8 Estimate the relative enthalpy of fusion of the following compounds, first assuming complete atomic mixing in the liquid and then assuming retention of

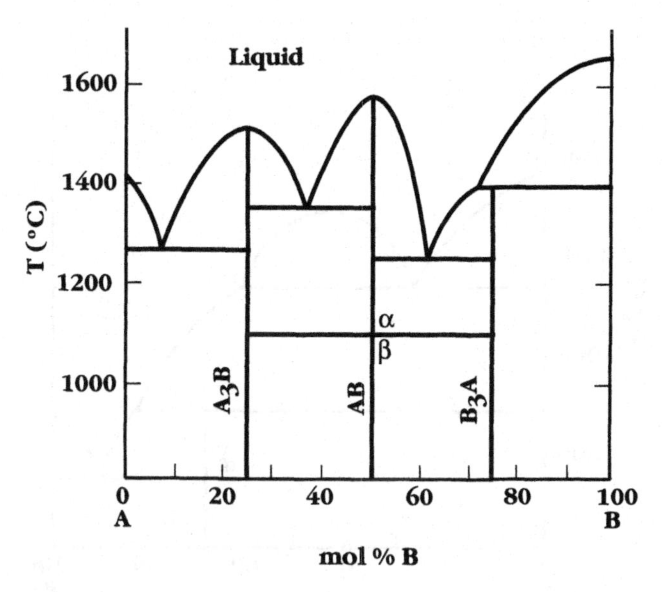

Exercise 6.6

simple stoichiometric single oxide groups in the liquid phase. (Assume similar melting points.)

$A^{6+}B_2O_7$ A_2O_3 $A^{3+}B^{3+}O_3$ $A^{2+}B_2O_4$

6.9(a) Label all regions in the attached binary diagram. Indicate common crystallization regions on the diagram. Write out the critical reactions (with estimated temperature) for each of the crystallization types.

(b) Assume that the chemical potential of A in B is exponential at small A and linear with composition at larger A content. Sketch a diagram of the variation of the chemical potential of A in the equilibrium phases across this system at 800°C. Is it possible to do this without the assumption of some solid solution of the endmembers?

6.10(a) For the system 30A:70B of the hypothetical phase diagram, sketch a hypothetical microstructure as resulting from homogeneous nucleation from the melt at $T_{liquidus}(-)$, followed by random attachment for phase growth. (Hint: start with a 10 × 10 or finer grid and use a random number generator to decide the location of the next "unit" of crystallization, using 50° temperature steps between lever rule calculation. Decide on rules of valid growth or attachment.)

(b) How does your microstructure change if heterogeneous nucleation only occurs, i.e., at the container walls?

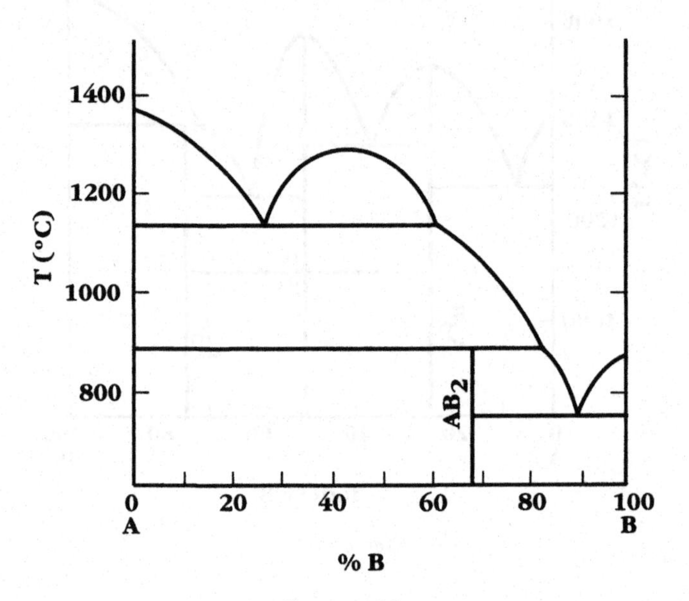

Exericse 6.9

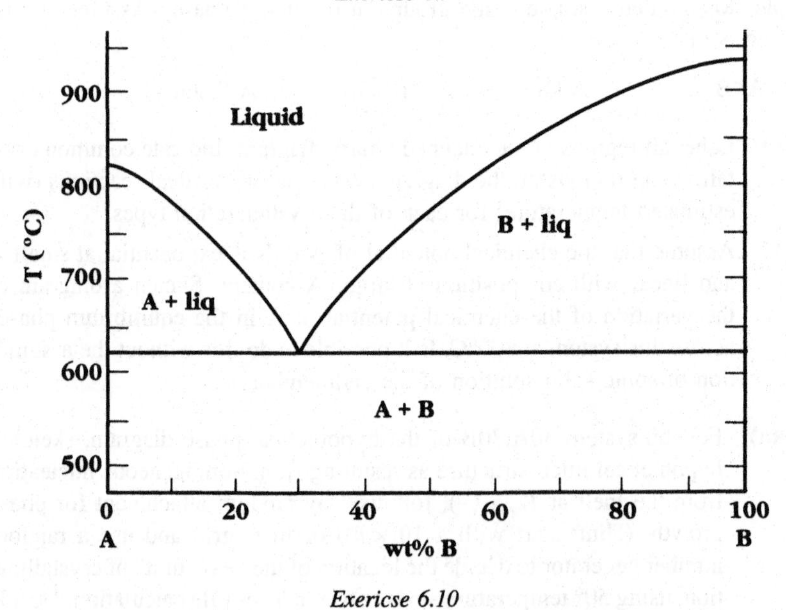

Exericse 6.10

(c) Did any melt areas become isolated during this simulation? What would occur in a real system in that occurrence?

6.11 Raw materials A and B (60 mol% B) are reacted together. The product microstructure consists of residual "A" coated with an AB outer layer surrounded by discrete AB_3 particles.

(a) Estimate the onset temperature for reaction.

(b) What does the observed phase assemblage tell you about the relative transport rates of A and B through AB?

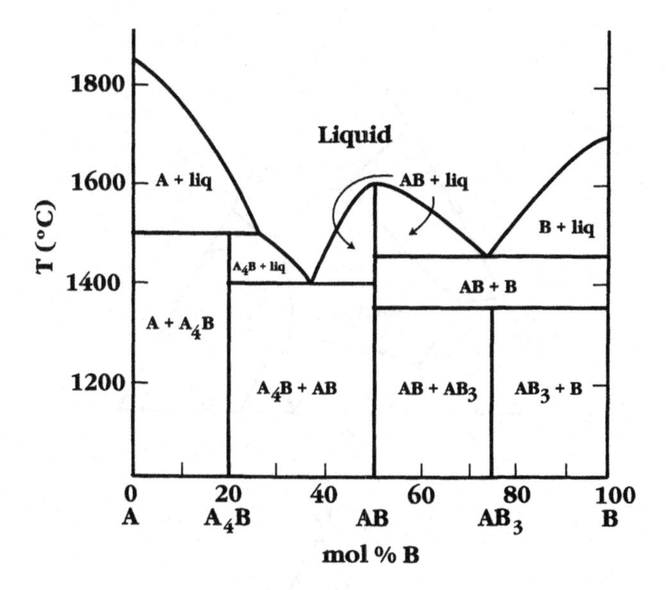

Exericse 6.11

6.12 Sketch valid phase diagrams for the binary systems described.

(a) The system A–B contains two congruently melting compounds, A_2B and AB_2, melting at 1200°C and 1400°C, respectively. The endmembers' melting points are A, 1100°C; and B, 1600°C. There are no measurable ranges of solid solution in either of the compounds or in either endmember. Justify your indicated eutectic temperatures.

(b) In the system K–L, there is complete solid solution between endmembers, which melt at 1200°C and 1800°C, respectively. How would this diagram change if the solid solution had a tendency toward separation?

Chapter 7

7.1 For the schematic diagrams, construct Alkemade lines and determine the crystallization map. Label binary and ternary peritectic and eutectic invariant points.

7.2(a) Draw Alkemade line(s) and determine the relative direction of decreasing temperature ("slope") of each boundary. Label peritectic and eutectic

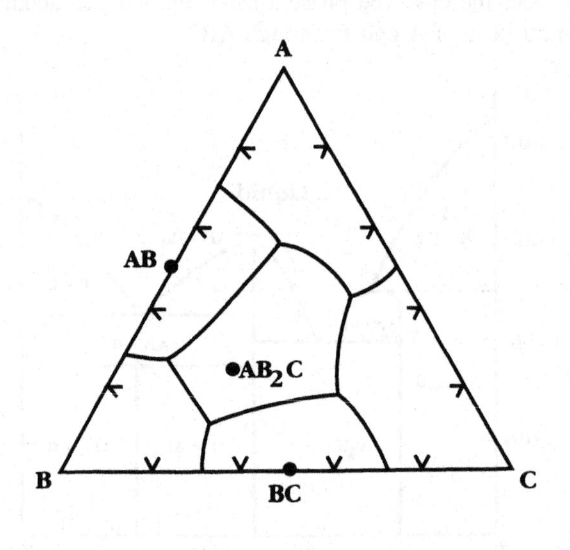

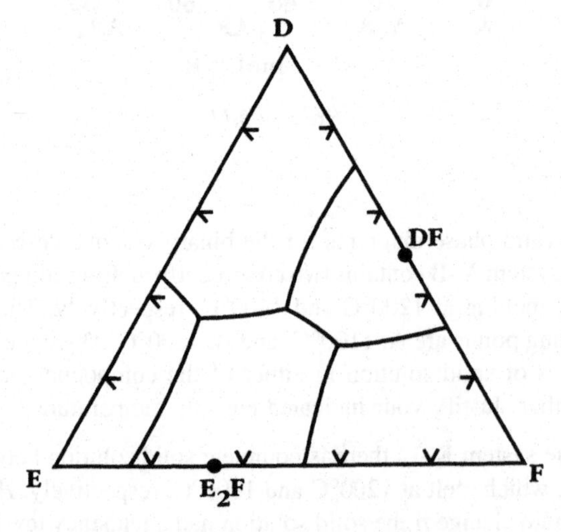

Exericse 7.1

points in the ternary and the side binary systems. Determine regions of common crystallization type.

(b) Locate the "reaction boundary" and highlight the two regions in which resorption of primary crystals will occur. Describe the crystallization of regions exhibiting resorption reaction.

(c) Draw isothermal sections for each temperature shown (700 to 1200°C). Indicate tie lines in two phase areas. Label three-phase areas with compatible phases.

(d) Determine the composition and describe the crystallization of composition points R and S.

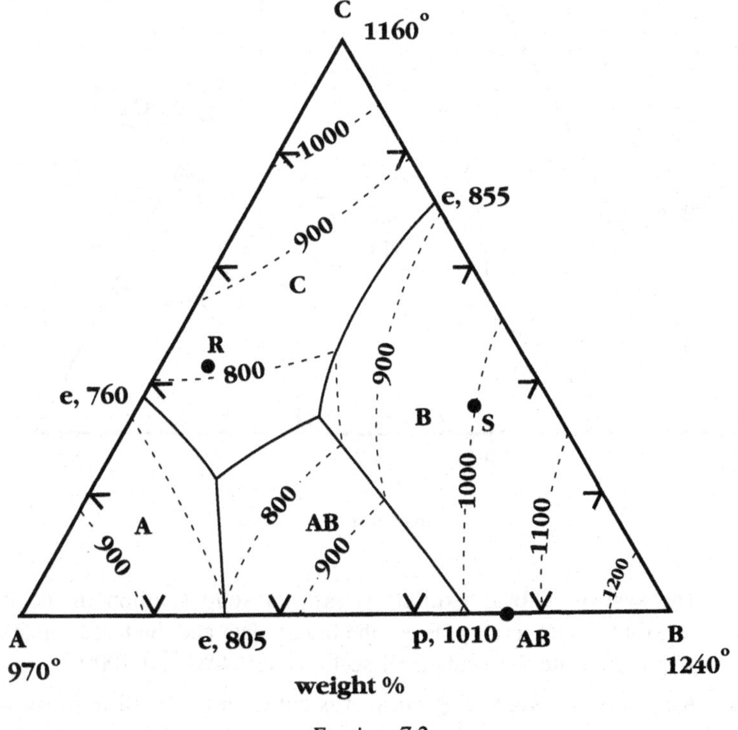

Exericse 7.2

7.3(a) Locate the composition 34% AO_2:24%BO_2:42%CO_2. In what compatibility triangle does the composition lie? What are the equilibrium proportions of the compatible phases?

(b) Given raw materials ACO_4, B_2CO_6, and AB_2O_6, use the phase diagram to determine the batch composition that would yield the desired composition of part A.

(c) Note that this diagram is given as mol%, but batch calculation must be stated in wt%. What does the "mass balance" of the composition point in the triangle defined by the raw materials signify?

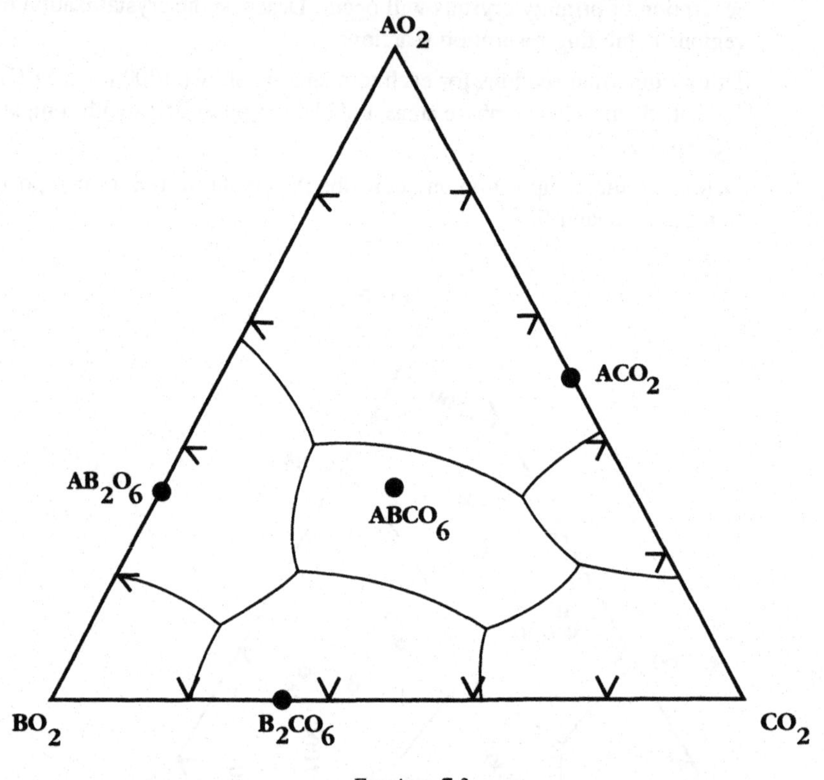

Exericse 7.3

7.4(a) The system A–B–C exhibits significant solid solution in the binary subsystems and ternary. From the binary data and the ternary isotherms given, generate the isothermal sections at 900°C and 700°C.

(b) Assemble a liquidus diagram that is consistent with all data given.

(c) Assemble a solidus diagram that is consistent with all data given.

Chapter 8

8.1 These questions refer to the phase diagram of the system SiO_2–MgO–Al_2O_3.

(a) A mixture of 30 wt% alumina (Al_2O_3), 50 wt% forsterite ($2MgO{\cdot}SiO_2$), and 20 wt% SiO_2 is prepared. Locate the composition and determine the equilibrium phase proportions.

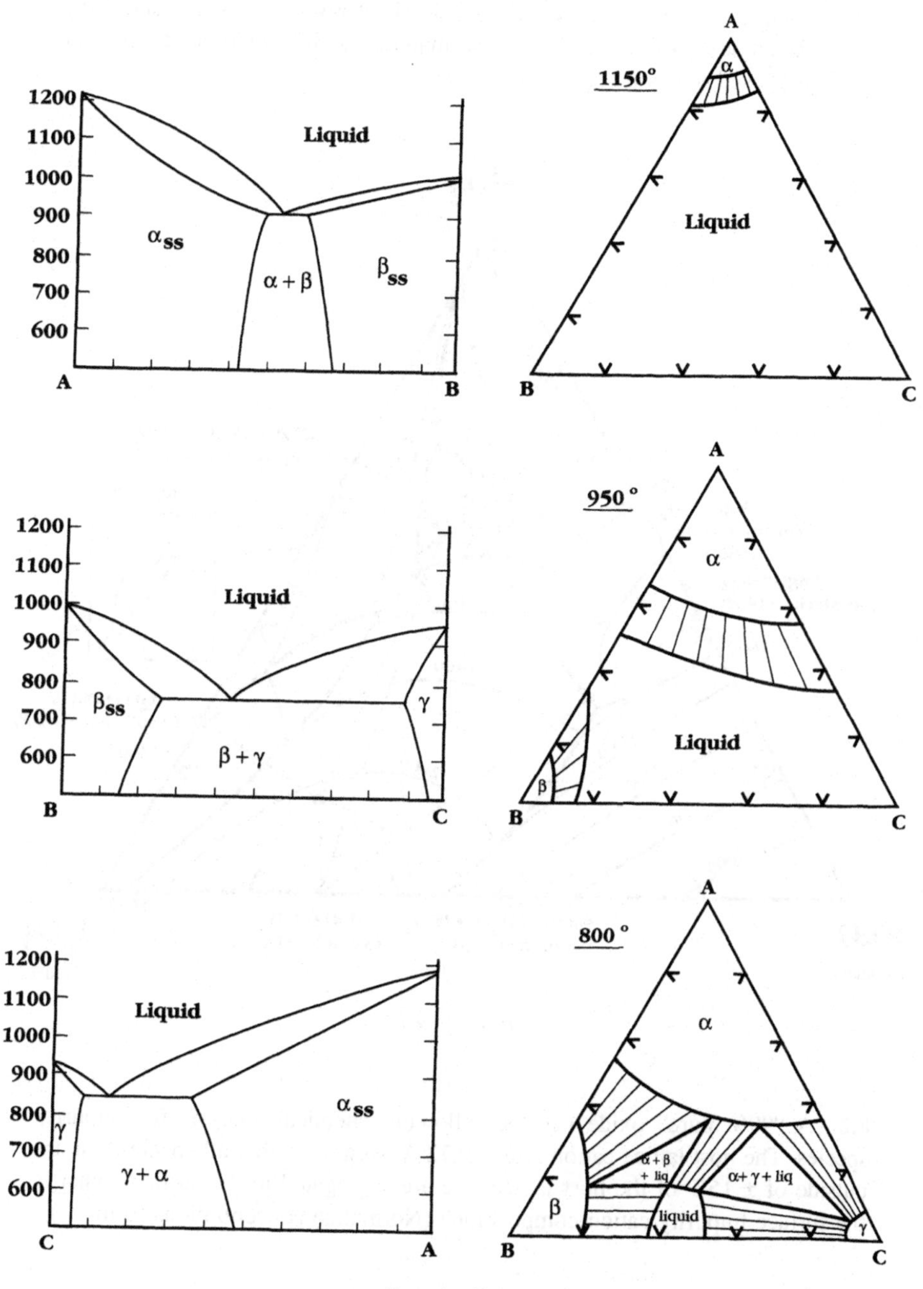
1200
1100
1000
900
800
700
600
Liquid
α_{ss}
$\alpha + \beta$
β_{ss}
A
B
1150°
A
α
Liquid
B
C
1200
1100
1000
900
800
700
600
Liquid
β_{ss}
γ
$\beta + \gamma$
B
C
950°
A
α
Liquid
β
B
C
1200
1100
1000
900
800
700
600
Liquid
γ
$\gamma + \alpha$
α_{ss}
C
A
800°
A
α
α + β + liq
α+ γ + liq
β
liquid
γ
B
C

Exercise 7.4

(b) Suggest the first reaction temperature and sequence of reactions during firing for all similarly small particle sized raw materials. Discuss the difference particle size and raw material purity could have on your proposed reaction sequence.

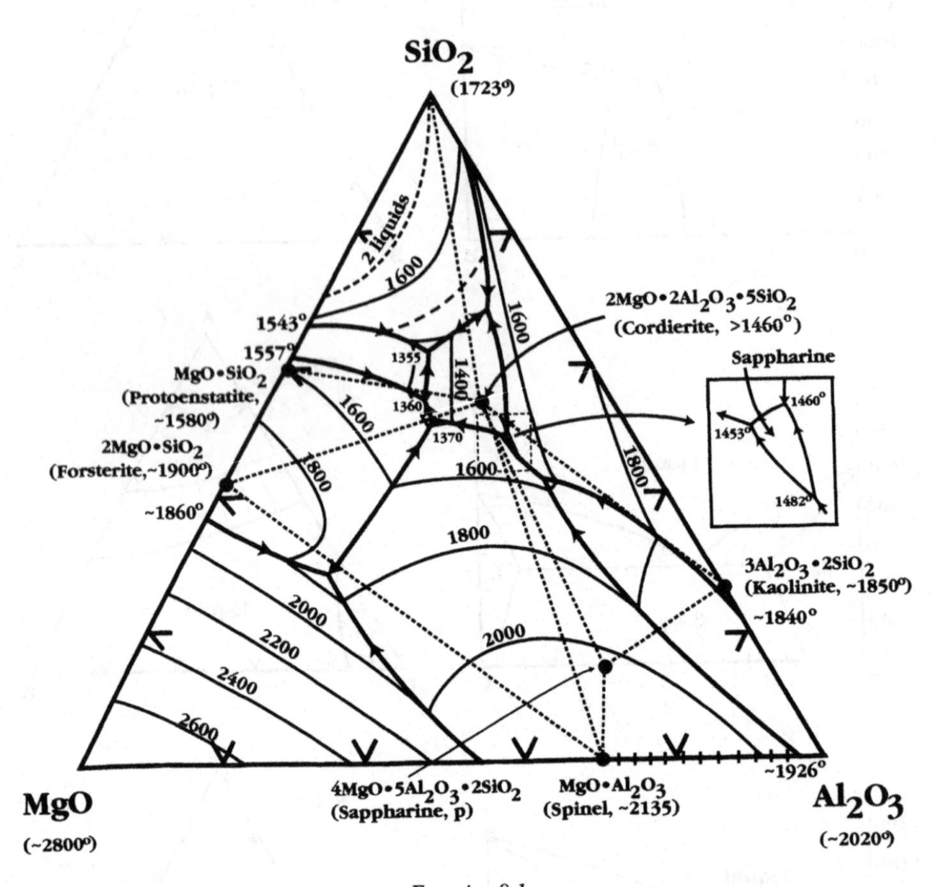

Exercise 8.1

8.2 A "99% pure" oxide has the following chemical analysis for impurity species. The host lattice cation size is 0.72 Å. Assuming that all species that are outside of ± 15% of the host lattice size are segregated to the surface, what is the surface impurity cation composition? (Normalize concentrations to unity.)

Impurity oxide	wt%	Six-fold cation radius (Å)*
Li_2O	0.05	0.76
K_2O	0.05	1.38
Na_2O	0.10	1.02
Al_2O_3	0.24	0.54
Fe_2O_3	0.32	0.6
SiO_2	0.20	0.40
TiO_2	0.05	0.61

*From R. D. Shannon (1976). *Acta Cryst.* **A32**:751–67]

8.3 A "forsterite" refractory brick has a nominal composition

CaO	1
MgO	55
Al_2O_3	1
Fe_2O_3	0.9
Cr_2O_3	1
SiO_2	34

(a) Using pH_o table values, determine the most likely compound forming pairs.

(b) Group similar species and speculate on solid solution formation.

(c) What compounds would you expect to find in this refractory? (Check your answer if possible.)

8.4 A current refractory wall installation exhibits a maximum erosion rate of 1 in./year in liquid contact at 1400°C. The current flow rate is 1 m/hour. It is known that the melt viscosity decreases about 10% per 50°C increase in this temperature range.

(a) Estimate the relative effects of an increase in melt temperature of 30°C versus an increase of flow rate of 10%. (Assume all other factors are unchanged.)

(b) The temperature increase will similarly increase diffusivities in the melt, altering the boundary layer thickness. Will this have an effect on the erosion rate? Estimate an effect from the temperature increase on solubility in the melt and consider the consequences on erosion rate.

Index